Living Roots

Living Roots:
The Promise of Perennial Foods

Edited by

Liz Carlisle and Aubrey Streit Krug

Illustrations by
Emmy Lingscheit

ISLAND PRESS IMPRINT
PRINCETON UNIVERSITY PRESS
PRINCETON AND OXFORD

Published by Princeton University Press
41 William Street, Princeton, New Jersey 08540
99 Banbury Road, Oxford OX2 6JX

press.princeton.edu

Since 1984, the non-profit organization Island Press has been stimulating, shaping, and communicating ideas that are essential for solving environmental problems worldwide. In 2026, Island Press became an imprint of Princeton University Press to further advance its mission and publishing.

GPSR Authorized Representative: Easy Access System Europe - Mustamäe tee 50, 10621 Tallinn, Estonia, gpsr.requests@easproject.com

All Rights Reserved
ISBN (pbk.) 9781642833881
ISBN (ebook) 9781642833898
Library of Congress Control Number: 2025947182

British Library Cataloging-in-Publication Data is available

Editorial: Emily Turner
Production Editorial: Sharis Simonian
Text Design: Sztrecska Publishing
Cover Design: Bruce Gore
Production: Danielle Amatucci
Copyeditor: Carol Peschke
Cover art: Mark Han, background courtesy of Shutterstock

This book has been composed in Adobe Garamond Pro and Myriad Pro

Printed in the United States of America

10 9 8 7 6 5 4 3 2 1

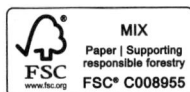

MIX
Paper | Supporting
responsible forestry
FSC
www.fsc.org   FSC® C008955

Contents

Introduction

There's something particularly lonely about attending to the minutiae of daily life while the planet is burning. You take your child to day care. The ocean warms to unprecedented temperatures. You drive to work. Your city sets another heat record. It feels like everything that really matters is somehow out of reach.

But all around us are fellow beings whose very metabolic nature proclaims their ongoing allegiance to the living world. Mussels filtering water. Bees pollinating wildflowers. Sheltering trees bearing fruit. This begs the question, Why can't we humans sustain ourselves in ways that sustain others too?

Right now, our food systems often function like the agricultural equivalent of fast fashion. Plants are only in the ground for this season. A handful of these annuals—corn, soy, wheat, and the like—dominate the landscape. Commodity crops are shipped out in tremendous volume, harvested and processed by people whose work puts their health and livelihood at risk. These crops are hurriedly assembled into cheap, low-quality products. Fossil fuels are squandered at every stage of the process, and it all ends in a stupefying amount of waste. Indeed, our food systems are currently responsible for a third of global greenhouse gas emissions, from the carbon dioxide lost when clearing land to the methane emitted from all that wasted food in the landfill.

As environmental scholars and educators, we found ourselves wondering, How do we build a food system that is made to last?

In both of our places, we started to learn about the diversity of food systems that flourished a few hundred years ago. In coastal California, where Liz lives and works, acorns featured as a staple food, alongside holly leaf cherries and elderberries. Chumash and other Indigenous peoples tended these prolific food forests with fire and an epic collective pruning effort. In the central Great Plains, where Aubrey lives and works, grasslands have been stewarded by the Kaw, Pawnee, Osage, Omaha, and other Indigenous nations. Prairies fed bison, elk, rabbits, prairie chickens, and the many people who relied on these creatures along with plums, berries, and a bevy of tasty roots.

The key to both of these food systems is that they worked with, rather than against, the creativity of the plants that evolved to live most abundantly across Earth's landscapes: perennials.

~

Perennials, roughly speaking, are plants that live for more than one year. You're probably familiar with many perennial plants already: Trees and bushes are perennial, as are the long-lived grasses you might see out in the rural parts of the country where we are from. You might have even eaten some perennial foods today: Tree nuts and tree fruits are obvious ones, but grass-finished meat raised on perennial pasture might be a candidate for this designation as well. What these foods all have in common is that they come from farms where plants stay in the ground for multiple seasons. Which means their roots can go to work.

Perennials stay in place for decades—and sometimes centuries—precisely because their roots are so active. Roots seek and exchange; they push and pull, tense and release; they brace, rub, infiltrate; they decay and turn over. Roots go deep, sometimes dozens of meters. Critically, they take sunlight energy harvested by the plants' leaves and deliver it into the ground to provide food for microbes, with which they form a sort of belowground superorganism. Microbes gather nutrients and water needed by the plants, while plants feed microbes the sugars they need to survive. Year after year, perennial plants allocate a quarter to a third of their solar harvest to their microbial mutualists, pulling carbon down out of the atmosphere and building up the terrestrial life support system commonly known as soil.

This rooting movement of matter and energy happens in the dark, mostly out of sight. But it is integral to how sunlight powers precious cycles of life.

~

You may know this story: An empathic teenage girl convenes a community and navigates a path forward amid water scarcity, political crisis, economic collapse, and acute climate change. It's the story of Lauren Olamina, the main character of Octavia Butler's prescient and enduring science fiction classic *The Parable of the Sower*, set in California in the 2020s.

In her journey, Lauren relies on seeds. Seeds as inspiration for her worldview and seeds as literal food stocked in her survival pack. Eventually Lauren and her community make it to a place they name Acorn, where they bury their dead and plant oak trees. In the face of apocalyptic devastation, Lauren's dreams of future life are made possible by her alliance with these deep-rooted old friends who have been with humanity through thick and thin.

Now more than ever, humans need both perennial foods and the living roots required to grow them. At this advanced hour, perennial plants can't avert the climate crisis. People must slow and stop the burning of fossil fuels. But where and how we grow perennial foods will help shape what human futures are possible. Perennials can help us significantly mitigate emissions, so the climate crisis doesn't get worse. They can provision a growing buffet of foods and medicines that help us navigate hard realities and limits, and they can supply meaningful relationships that inspire us to create better place-based communities. These hardy survivors can help us weather the heat, drought, and storms that are coming our way with increasing frequency.

Perennials play the long game, spreading the wealth both above and below ground to find a durable balance. This is how they deal with carbon. Whereas annual plants like wheat and corn allocate only 15 to 30 percent of their total carbon stores underground, native perennial grasses and forbs allocate 50 to 67 percent of fixed carbon in root or rootlike tissues below the soil surface. Conversion of these perennial landscapes to annual cropland has resulted in 35 petagrams of carbon escaping to the atmosphere, roughly equivalent to three to four years' worth of worldwide emissions from fossil fuels and cement. Just imagine if we began to bring these perennial ecosystems back.

Likewise, nitrogen has historically been carefully stewarded by perennial plants. This element gets less press than carbon, since carbon dioxide is so much more prevalent in the atmosphere than its nitrogenous greenhouse gas cousin. But nitrous oxide has 300 times the warming power of its more famous fellow greenhouse gas, so it's worth paying attention to. Some 55 percent of it comes from agriculture, due to overuse of synthetic fertilizer and fields being left bare for much of the year. All this excess nitrogen from crop fields also escapes into the watershed, which is why agriculture is this country's largest source of water pollution and there are over four hundred marine dead zones around the globe.

The good news is, perennial plants can help us clean up this mess. By restoring them where they belong, we can replace at least some of the carbon lost from these landscapes, drawing it back down from the atmosphere. Perennial plants don't need to be replanted each year, so they can alleviate the need for tillage and instead hold soil and its carbon in place. They can be grown with little or no fossil fuel–based fertilizer. And they can scavenge much of the nitrogen currently running off farm fields, intercepting potential pollution and recycling it into food for microbes—and eventually people too.

⁓

But in order to revive these critical climate allies when humanity needs them most, we will need to reciprocate, to come to the aid of perennial plants as they have come to ours. We will need to restore rightful land access for Indigenous communities who have long stewarded plant relatives. Those of us who have been disconnected from land for generations must take care not to appropriate the traditions of Indigenous communities but instead rekindle our own friendships with photosynthesizers. It will take all of us working in concert, which is where you come in. It might start with a new recipe or a new sapling tucked into your backyard. You and your neighbors might plant a food forest or volunteer to help a local farmer establish a hedgerow. One way or another, we can all nurture the living roots of primarily perennial food systems.

Humans stand on top of the ground; unlike roots, we can bask in the sun. But our lives are interdependent with our fellow beings, with perennial plants whose trunks and stems can tower over us, whose rooting depths can outreach us. And like roots we live in the dark when it comes to knowing much about the future. What we do know is our powerful propensity for seeking, making, learning. Our technology and labor—not to mention other animals and tools—are woven into the fate of our homeland on Earth. What we do now matters.

The tricky beauty of our reality is that as a species we know how to do many things. Many of us live in societies built through the burning of fossil fuels, but selecting seeds, pruning trees, and singing songs are human enterprises as well. People are constantly solving problems and getting ourselves into trouble. We know how to fertilize wheat and breed perennial grains, scroll social media and pluck serviceberries, drive cars and run long distances. We wage wars, but we can also extend olive branches. Perhaps we must now discern which of these activities can endure, tell the truth about which of these we want to persist and grow. Are we willing to try to live? What gifts do we each have to bring for the benefit of all?

~

From the earliest days of civilization, many communities developed intimate knowledge about the perennial plants they harvested to feed their families, fuel their hearths, and weave the tools of everyday living. Eighty thousand years ago, people in what is now southern China were likely eating carbohydrate-rich nuts as part of a varied diet. Sixty-five thousand years ago, people in what is now Australia were grinding the grainy seeds of native perennial grasses. These plants helped people develop ways of life that fit their places, rich with stories and skills passed down to children and grandchildren.

Perennial foods have been the cornerstones of cuisines and cultures. Remember acorns, aptly highlighted in *Parable of the Sower* as a staple food for many peoples around the world. Or the nutrient-dense quartet of olives, dates, chestnuts, and carob, which have nourished people of the Mediterranean basin for millennia. Walnuts and pistachios in Kyrgyzstan. Oil palms in the tropics. Camas bulbs in western North America. Protein-rich and drought-tolerant mesquite in the Mesoamerican desert.

And yet, today, some 60 to 80 percent of global cropland is dedicated to annuals, just four of which (corn, wheat, rice, and soybeans) account for 75 percent of the calories consumed by people.

There are deep-seated reasons for this dramatic shift toward a handful of annual foods, entrenched political and economic forces that have pushed agriculture toward short-term profits rather than long-term human and environmental health. Building robust perennial food systems means challenging these forces and the corporations and decades of received wisdom arrayed behind them. We'll also need to repair the damage already wrought by extractive annual agriculture, regenerating soils and social bonds so we can sustain people and the land. And we'll need to build systems to ensure that the ecological and economic benefits of more perennial landscapes will be fairly distributed. A bold movement into a more just, perennial food future is no small task. It will require badgering our elected officials, having courageous conversations with our neighbors, and embracing new flavors and rural aesthetics.

And yet we think people are up for it. Already, in fact, many thousands of people are standing up for perennials and the lifeways that can go with them. In this book, you'll hear from Indigenous scientists and community leaders who are working to restore buffalo prairies and traditions of berry gathering. You'll also hear from urban visionaries planting food forests. Farmers planting fruit and nut trees between their crops and hedgerows at the edges of their fields. Ranchers learning to graze their livestock in patterns that mimic the behavior of native herbivores, so as to steward healthier grasslands. And you'll hear from scientists and farmers who are developing and stewarding perennial grains, from sorghum to silphium.

These efforts are wildly diverse, much like a healthy forest or prairie. People don't all have the same life histories and strategies. And yet we do share a common home planet, and we will have to work with perennial plants to care for it.

We, the contributors to this book, are willing to devote our lives to this work, to realizing new and renewed access to perennial foods and landscapes. We share a vision for a food system that can sustain us and our descendants and their descendants and their descendants, for generations into the future. A food system that doesn't leave anyone out, that relieves the climate chaos and the burden of anyone with polluted water or exploitive working conditions. A food system in which both plants and people nurture the sustaining collective power of living roots.

Some of the stillness I have found is bound to last
Some of the restlessness will live on
Some of the pain I've always known is hard to pass
But a mutual blessing takes the game on

Summer seed become my perennial bloom
Summer's healing coming soon

—Lukas Nelson & Promise of the Real
(lyrics from "Perennial Bloom (Back to You)")

Wildness. . . . It is perennially within us, dormant as a hard-shelled seed, awaiting the fire or flood that awakes it again.

—Gary Snyder
(from *The Practice of the Wild: Essays*)

the profound change
has come upon them: rooted, they
grip down and begin to awaken

—William Carlos Williams
(from "Spring and All")

It is only when we are fully rooted that we are really able to move.

—Madeleine L'Engle
(from *A Wind in the Door*)

I. FOREST

OFTEN CALLED THE LUNGS OF THE EARTH, forests cover more than 30 percent of the planet's land surface, producing air for the rest of us to breathe. Tropical forests alone are home to two thirds of all terrestrial plants and animals. As magnificent as these ecosystems are aboveground, they are equally impressive belowground, where vast networks of fungal mycelia link seemingly distinct trees into an interconnected superorganism.

Despite ongoing deforestation—as well as recent pressure from increasing temperatures, drought, wildfire, and new insect pests—forests absorb more than a quarter of human carbon emissions each year. Stopping deforestation and restoring these unique ecosystems, scientists suggest, could offset even more: up to a third of current carbon emissions worldwide.

People living in forested parts of the world have long created food systems with and among trees. Mayan communities are famous for cultivating elaborate gardens in the lowland forests of modern day Mexico and Central America, sowing root crops and squash on the forest floor, corn and beans above these, and fruit trees and bushes in an intermediate layer below the canopy of palms and hardwoods. Forest gardeners across the African savanna and tropics are likewise renowned for multilayered intercrops of grains, vegetables, and trees, a farming method that "transforms a rainforest into a food forest."

Such methods, known as forest farming, are the longest-lived form of agroforestry, the integration of trees and shrubs into agriculture. Other agroforestry practices include alternating tree crops and row crops (known as alley cropping) and raising livestock under the canopy of trees (silvopasture). Farmers also plant trees and shrubs for habitat and soil health (hedgerows), for shelter (windbreaks), or to protect the water quality of streams (riparian buffers).

Over the course of the past century, agriculture and trees have had an uneasy relationship with each other. As tree cover has been removed around the world to make way for row crops and plantations, catastrophic emissions and loss of wildlife habitat have followed. Undoubtedly, deforestation remains a serious and unresolved problem, one of the most daunting and important fronts in the battle against climate change. And yet a small but increasing number of farmers are actually restoring trees to the landscape, spawning a bevy of new agroforestry initiatives and research projects. The idea that trees and farms belong together is gaining ground, and with good reason.

The climate case for agroforestry is a powerful one. A global synthesis of research on the subject found that soil organic carbon increased 40 percent in the top foot of soil when land transitioned from annual agriculture to agroforestry. Another study suggests that widespread incorporation of trees and shrubs into North American farms could for a period of time sequester enough carbon to offset 34 percent of US annual emissions from coal, oil, and gas. And that's just the carbon. Research conducted in Illinois found that recently planted alley cropping systems reduced annual nitrous oxide fluxes by 25 to 83 percent over the course of four years compared with adjacent row crops, and a Canadian study found 74 percent lower nitrous oxide emissions in Saskatchewan shelterbelts. Accounting for the combined reductions in these two powerful greenhouse gases, climate think tank Project Drawdown ranked three agroforestry practices (silvopasture, tropical staple trees, and tree intercropping) in its top twenty climate solutions across all sectors, well ahead of electric vehicles.

Trees and shrubs also help farmers ride out rising temperatures and extreme weather. Perhaps most obvious to anyone who's ever been outside on a hot day, trees provide critical shade for people, livestock, and crops that need a little relief from the heat. Trees stabilize temperatures belowground as well, in addition to protecting crops from wind and holding water.

Beyond their climate benefits, trees can improve rural environments in lots of other ways too. They greatly reduce water pollution from fertilizer, providing a safety net to catch nitrogen before it leaves a farm field. They also provide crucial habitat for all sorts of species that have trouble navigating landscapes like the US corn belt, notably birds and pollinators. Humans, too, benefit from trees: A recent analysis of health benefits from exposure to trees found a lower incidence of respiratory problems, lower exposure to ultraviolet radiation, reduced risks of heat-related illness and death, improved mental health, less stress, and better immune function and cardiovascular health.

All this and it might be better for farmers' bottom lines too! One study found that perennial crop income averaged nearly three times annual crop income over the same ten-year period. There are many reasons for this, but one is that growing trees and crops together can make more efficient use of light and space, meaning more crops per acre.

So far, though, farmers interested in agroforestry have been largely on their own. Hazelnuts and elderberries don't enjoy the same subsidies that make corn and soybeans such a sure bet, and most agricultural advisors aren't familiar enough with tree crops to help farmers establish them successfully. Markets for many of these crops are in their infancy, which means that finding a nursery to sell you plants or a processor to grind your chestnuts into flour can be difficult to impossible. It's no surprise then, that agroforestry remains rare on US farms. The 2017 Census of Agriculture used a very liberal definition of agroforestry, where windbreaks alone would qualify. Still, just 1.5 percent of US farmers reported using at least one agroforestry practice.

But things started looking up for agroforestry with the 2022 passage of historic climate change legislation, which included millions of dollars in grant funding for agroforestry projects across the country. Although this one-time funding hasn't changed the fundamental structure of US agricultural policy, which remains strongly biased toward commodity row crops, it has demonstrated what's possible when incentives and technical assistance for tree crops become competitive. The US Department of Agriculture conducted its first national survey of agroforestry producers in 2022, signaling the sector's growing importance.

Another way of thinking about the potential for agroforestry to grow is to take stock of all the land currently in row crops that may be more suitable for trees or shrubs. As droughts and floods intensify, planting corn and soy on hilly or low-lying ground becomes a riskier and riskier proposition. Where farmers find themselves frustrated by chronic soil erosion or flooding, they might save money by converting these areas to hedgerows or orchards instead. A group of researchers in Missouri recently mapped marginal lands in their state that might be suitable for conversion to nut tree production, identifying more than 6,000 square miles suited for black walnuts, more than 4,000 square miles primed for pecans, and nearly 1,800 square miles that might be planted with Chinese chestnuts.

Ultimately, whether they are planted in an orchard, in between farm fields, or in urban gardens, the collective benefits that trees and shrubs provide can add up. With each new planting, an entire watershed becomes a little more resilient, bolstered for the next heavy rain. And as tree cover on farms continues to grow, these woody perennials push agriculture ever further away from its recent past as a carbon source toward a future as a carbon sink.

Throw It Upwards

Leah Penniman

Hyacinth Laratte sunk her pickaxe into the dusty crust of soil on the mountain slope of Cormier, outside of Léogâne, Haiti. The earth yielded, and I tipped the tattered, repurposed jug of water toward the depression, splashing just enough water to darken the earth. Deftly, Hyacinth plunged a vetiver grass seedling into the moist hole, entrusting it with an ambitious and crucial responsibility: to hold the mountain soils in place.

In the Haitian countryside, farmers will tell you, "Tè a fatige," which means, "The earth is tired." The earth's exhaustion began in the seventeenth and eighteenth centuries, when the French empire cleared virgin old growth forest to make way for sugar plantations to be toiled by enslaved African people. After the enslaved farmers fought for and won their independence against staggering odds, their former enslavers demanded an independence ransom that cost Haiti an estimated $115 billion in lost economic growth. To pay this debt, Haitians exported mahogany to France, to the tune of 4 million cubic feet annually in the mid-1800s. The denuded slopes no longer stabilized the soil, and the fragile earth tumbled into the sea. The direst of studies report that Haiti retained less than 1 percent of its original forests.

Hyacinth is among the generations of Haitian land workers dedicated to restoring the forested slopes of this land of mountains beyond mountains. They sow a densely rooted perennial grass called vetiver in contour strips on the dramatically steep hillsides. Once the vetiver is established, it holds the soil in place, and the farmers can plant mapou, acacia, mango, moringa, and cherry trees in the stable soil behind it. Approximately every 20 feet, they dig 1-meter-deep trenches along the contour to catch runoff and soil that inevitably washes down the hillside during the heavy spring rains. Periodically, these trenches are dug out and the soil returned to the planting mounds. The combination of vegetated strips and soil conservation trenches increases water infiltration and maintains soil health. The efforts of Haiti's rural land stewards have reaped rewards, and today Haiti's forest cover is around 32 percent.

Each winter after the devastating 2010 earthquake, I joined a diasporic delegation of skilled volunteers working with Hyacinth's community on reforestation efforts. Together, we planted thousands of trees in contour strips and witnessed the fragile dust transform into humus-rich soil. Inspired by the ingenious methods of our Haitian friends, I wondered whether similar techniques could repair the marginal slopes on our newly acquired farmland in the original Mohican homelands of Grafton, New York.

I remember the first time I plunged a shovel into Grafton's mountainside soil that would become Soul Fire Farm. I was able to dig precisely 7 inches before hitting hard gray clay, virtually impenetrable to crop roots and resistant to even the metal blade of the tool. In contrast, the prime river bottom soils of the Midwest have a typical topsoil depth of more than 40 inches. The open, south-facing portion of our land was on a steep slope, dangerous to navigate by tractor and highly susceptible to erosion. In fact, we discovered that much of the topsoil had washed down the slope over the years and was trapped on the uphill side of the stone wall at the property's edge. The New York State Department of Agriculture and Markets ranks agricultural soils by their potential to yield nourishing crops, and our soils were listed near the bottom as "marginal, unsuitable."

As for so many Black farmers throughout time, marginal land was all that was within reach for our family, and the only option was to work with it. Like our Haitian friends, our family planted contour strips of fast-growing, tenaciously rooted herbaceous plants and woody shrubs on our slopes: comfrey, horseradish, mint, yarrow, chives, elderberry, and willow. We fortified the planting strips with punky logs harvested from the forest and backfilled with soil that we rescued from the bottom of the slope. We later learned that in Ki-Swahili, this practice is called *fanya-juu*, meaning "throw it upwards." East African farmers revive degraded land by recovering soil from the bottom of the slope and throwing it upward to form terraces, or steps of roughly level land. The practice of *fanya-juu* results in soil organic matter levels 35 percent higher than that of conventional farms, comparable to natural ecosystems. This practice also results in 25 percent higher crop yields than conventional farms. Atmospheric carbon dioxide is sequestered in the terraces, mitigating climate change.

Once our *fanya-juu* earth-moving process was complete, we planted trees and additional shrubs in the new terraces: apple, plum, peach, hazelnut, apricot, pear, honeyberry, currant, jostaberry, and serviceberry. Each woody plant was tucked into a guild of beneficial herbal friends: daffodil as pest repellant, bee balm as pollinator attractor, garlic to ward off rodents, and comfrey as the supreme chop-and-drop mulcher. We heaped generous mounds of our homemade wood chips on the biodiverse collage of our terraces, inoculated with spent fungal blocks from the local mushroom farmer. As instructed by our elders in Ghana, we prayed over the plants, poured libations, sang, danced, made offerings, and gave thanks for the blessings of the abundant earth.

The results were miraculous. In just two seasons, our slopes transformed from denuded hardpan to a lush, self-sustaining food garden buzzing with hummingbirds and honeybees. It looked a lot like the established *jaden lakou* of Haiti, the polycultural house gardens of diverse trees, shrubs, agricultural crops, and livestock fodder that could be found in nearly every rural compound. We named this first half-acre experiment *Jaden Lakou* in honor of our Haitian friends who inspired it. My joy at the flourishing of our first agroforest earned me the enduring nickname "Perennials Papa."

This Perennials Papa, together with a growing team of family, friends, and farm staff, got to work expanding our tiny *jaden lakou* into a 4.5-acre diversified agroforest. The east field, tired out from years of annual crop production, begged to be included in our perennials experiment. In response, we planted multistory strips at 40-foot intervals across the entire field, interweaving fruit trees with elderberry and currant shrubs and undersowing a groundcover of wildflowers, aromatics, and medicinal herbs. In between the strips we planted a robust ruminant pasture mix of orchardgrass, clovers, ryegrass, field peas, chicory, birdsfoot trefoil, and alfalfa. Once these plants were established, we grazed our Nubian goats, Katahdin sheep, and laying hens in rotation in the alleys. Their manure fertilized the silvopasture and maintained a clipped and tidy pasture. When mature, the trees will also provide shade and dropped fruits for our herd and flock.

Our *jaden lakou* and silvopasture agroforests quickly became a centerpiece of educational programs at Soul Fire Farm, with the annual crop acreage taking a backseat. Annually, we welcome between 1,000 and 2,000 rising generation Black, Indigenous, and People of Color farmers to study agroecology with us. The polycultures are one of the outdoor classrooms where skills from tree planting to herbal medicine

Species in the Jaden Lakou Agroforest and Silvopasture
of Soul Fire Farm

Trees	Shrubs and Vines	Herbaceous Plants	Pasture
Apple	Aronia berry	Anise hyssop	Alfalfa
Apricot	Blackberry	Asparagus	Birdsfoot trefoil
Cherry	Blueberry, lowbush	Bee balm	.Chicory
Peach	Blueberry, highbush	Blue vervain	Clover, red
Pear	Chokecherry	Canadian ginger	Clover, white
Plum	Elderberry	Catnip	Orchardgrass
Willow	Grape, Concord	Chives	Pea, forage
	Hazelnut	Comfrey	Ryegrass, perennial
	Honeyberry	Daffodil	
	Jostaberry	Echinacea	
	Kiwi, hardy	Feverfew	
	Raspberry	Garlic	
	Serviceberry	Horseradish	
		Lavender	
		Lemon balm	
		Mint, peppermint	
		Mint, apple mint	
		Mugwort	
		Nettle	
		Oregano	
		Rhubarb	
		Rue	
		Sage, mountain	
		St. John's wort	
		Strawberry	
		Sweet Annie	
		Valerian	
		Yarrow	

making to livestock tending are practiced in a culturally connected and joyful manner. For example, during a recent weeklong Farming-in-Relationship-with-Earth (FIRE) immersion, we made a fresh batch of Indigenous microorganisms by harvesting healthy detritus from the forest floor and growing it in a medium of grain and molasses. The students squealed in delight, mixing the batch by hand, sweet molasses and aromatic humates oozing between fingers and tantalizing the senses. The fermentation process generated an inoculant teeming with beneficial bacteria, fungi, and nematodes, and we used that mixture in our holistic orchard spray to support tree health.

Soul Fire Farm's emphasis on agroforestry extends beyond our commitment to ecological health—it's about justice and cultural history as well. Our farm's mission is to uproot racism and seed sovereignty in the food system, working to heal our communities from generations of oppression and dispossession on land. Trees, and the soil they root in, have always mattered deeply to our people. Immediately after the end of chattel slavery in the United States, the Freedpeople of Falls Church, Virginia, wrote to the Freedmen's Bureau commissioner with this plea: "We feel it very important that we obtain HOMES, owning our own shelters, and the ground, that we may raise fruit trees, concerning which our children can say, 'These are ours.'" It is remarkable and stirring to take notice of the basal yearning for trees as part of freedom.

In a similar vein, Harriet Tubman, Moses of African American people, conductor of the Underground Railroad, felt that true freedom would be marked by the opportunity to plant her own trees for the benefit of her own community. She recalled that when she was enslaved as a child, she had been forbidden to eat the fruit of the trees she had been made to plant. She remarked, "I liked apples when I was young, and I said, someday I'll plant apples myself for other young folks to eat. I guess I done it." Indeed, Tubman planted an apple orchard at her house of freedom in Auburn, New York, which bears fruit to this day.

In a cruel and ironic twist, and despite generations of labor and yearning, Black Americans today own and operate very few orchards. In fact, only about 1.5 percent of all agricultural land is in Black hands. On the other hand, Black farmers are essential laborers on apple orchards in our state, traveling from Jamaica each season to harvest fruit over long hours and for low wages as part of the H2A guest worker program. This juxtaposition is not by choice but by design in an agricultural system devised for profit at the expense of workers, human communities, and the planet. Black, Indigenous, and Latine land stewards bear a particular burden as the survivors of genocide, chattel slavery, sharecropping, US Department of Agriculture discrimination, lynching, convict leasing, burning of towns, worker exploitation, and outright land theft. Healing our societal relationship to agriculture will necessarily include the return of lands, resources, and respect to these dispossessed farmers.

Alongside reparations, legislative victories, and land rematriation, an important part of restoring justice to the agricultural system will be uplifting and honoring the regenerative power of Afro-Indigenous farming techniques. On our once-degraded lands, we implemented the African diasporic technologies of *fanya-juu*, *jaden lakou*, and Haitian contour strips over the past fourteen years. We were eager to see how these heritage practices affected the earth, and so we invited Black soil scientist Tiffany LaShae to our farm to assess the underground health of our agroforest.

One sunny afternoon, Tiffany plunged her auger into our *jaden lakou* and pulled out sequentially deeper soil cores, expertly pointing out evidence of ancient glaciation and Indigenous-managed forest fires, then skipping ahead to the narrow black band that represented the sheep fever of the early 1800s. She proceeded in that manner, narrating the history of the land in colored bands of earth. And then she measured 10 inches of beautiful, dark, fragrant topsoil, proclaiming, "This was built in the last decade. You all helped create this soil. The geological record will bear your mark of healing!" We gasped in wonder! In a society obsessed with quarterly returns, we who believe in ecological repair must measure our contribution not in one-, five-, or even ten-year strategic plans. We will know that the words of our mouths, meditations of our hearts, and work of our hands will be acceptable in Mama Earth's sight when our contributions can be measured in geological time, etched into a record of stone.

Tree-Range Chicken

Reginaldo Haslett-Marroquin

When you visit my farm, it will probably take you a while to find the chickens.

The only time they are fully visible is early in the morning as they rush out of the coop. A cacophony of clucking and flapping ensues, as the early birds fly over the flock to get in front and reach the field feeders that we've placed strategically to help them expand their range. The noise and stampeding subside as they focus on breakfast. A few minutes later they start to look around, drink water . . . and disappear from sight.

Within a matter of minutes, you would be hard pressed to find a couple hundred of the 1,500 birds that make up every flock. They're hidden behind trunks and beneath branches, pecking on a multitude of plants and sprouted grains under the canopy of hazelnut trees. Their free ranging is possible only because of their confidence in the tree cover to give them shade and protection from predators.

And I think the trees know this, too. Hazelnuts have evolved to provide a thick and complete canopy: No sun gets through, and no hawk or eagle can see through either. I am convinced that thousands of years ago, nut trees and chickens made an agreement. One would provide shade and protection, and the other would provide a high-nitrogen fertilizer and control weeds and pests. It is no coincidence that the root system of a hazelnut tree is so massive, expanding up to 15 feet in diameter and down to 10 or 12 feet deep. These trees literally created the ideal system to process poultry manure, and the chickens evolved as "inefficient" feed converters, according to scientists who study the conversion of feed into meat for animals raised in confinement. However, we should remember that there's another kind of efficiency in the number of outputs that a circular, regenerative system generates.

A normal day on the farm is surprisingly quiet. No fighting, no stress. Even on 115°F days, at most the chickens pant while dissipating their body heat directly into the cool ground under the hazelnut trees. Incredibly, body heat from the chickens is also something the nut trees have adapted to. Above the hazelnuts, other tree species in this chicken habitat can include (in Minnesota) basswood, oaks, sugar maples, and hickory. And for a midlevel story we are introducing mulberries, another fruit tree that has also evolved with birds around it.

Why did the chicken cross my mind? Chickens are a jungle fowl, evolved in the rainforests of Southeast Asia. But in one of the more shameful chapters of human history, large-scale agribusinesses have confined these birds (and other animals) in massive, crowded warehouses. So my partners and I decided to bring chickens back into their forest habitat, or the closest modern version of it that we could envision. Instead of taking chickens away from their natural environment and bringing them into confinement—and then building a system to optimize the extraction of these chickens' value and the labor of those who care for them—we wanted to build a system around the needs of the chickens and the never-ending chain of positive impacts and outcomes that could come from recreating their habitat. When we designed the poultry production unit, integrating trees was just a question of how.

Chickens are also old friends of mine, since I come from the jungle too. I was born in eastern Guatemala and moved to the northern rainforest when I was four years old. We planted one of the few permaculture farms in that region. When everyone was slashing and burning the old-growth rainforest, we were planting a multispecies perennial farm with nine kinds of bananas and creamy round avocadoes far tastier than any I can find in the market today. The canopy of our farm also included oranges, mandarins, zapotes, and timber trees, and we raised shade coffee, pacayas, yams, yucca, and herbs in the understory.

During coffee harvest we had to battle the large chachalacas, or wild chickens, that went nuts over the sweet coffee cherries. They would come from the edge and work their way into the coffee rows. Armed with slingshots, we would wait downwind and score a few for dinner here and there. If we let them eat enough coffee berries, they went from skinny and hungry to plump and juicy. But we also learned that they would follow the corn kernels we left for bait and get trapped in baskets we installed with triggers they stepped on as they reached for the corn.

Our farm was a full one-hour walk from our house, so we stayed at the farm for weeks at a time when school was out of session. In my case, I spent a full year in the forest after finishing sixth grade. As far as my father was concerned, that was enough formal education for anyone. He did not write or read, so his knowledge came from observing, interacting, meditating, and interpreting what the ecosystem was so eager to teach us in the first place. Having come this far, as I write this essay, I smile at the amazing wisdom he carried and shared and at how right he was. He was so keen at observing and learning that he even predicted the invasion of our communities by the extractive industries that would see only quick profit from degenerating the landscape, undermining our food systems.

Back at home, closer to the school and town, we raised chickens under a similar canopy of oranges, guanabanas, nut trees, zapotes, and bananas. Our focus was on egg layers, and when they stopped laying they made the best soup anyone could wish for. We raised them completely outdoors, with no neighbors except jungle and grass fields, and I took charge of learning their behaviors so I could find their nests and perching trees. When I went back to school, I entered a 4-H–style competition to raise a rooster. Not only did I win first place, but I don't think I ever had a dinner that meant so much. At thirteen years of age, being responsible for raising the food—and then feeding my whole family a meal reserved for special occasions only—truly made me feel like a grown man. For days afterward, I walked around with the sense of pride typically reserved for the folks coming back from months of hunting in the forest to bring home much-needed products for the market and food for their families' tables.

So here I am in Minnesota, far away from the rainforest environment where I grew up in Guatemala. Yet I still carry the same dream of feeding my family and many others through Indigenous regenerative ways of working with the land. I envisioned a farm named after my mother's family, Salvatierra, which means Earth Saver. When I first moved to the upper Midwest, I visited the Native elders at White Earth and other places where the wisdom of ancient prairies and savannas is still respected. I was told that in the old days, hazelnuts were used as currency. This made sense to me, as I knew about how we used cocoa beans in ancient times in Mayan territories in Guatemala as currency. But there was a lot more. I learned that like humans, plants gather as communities, with different families living together. So I looked further at the hazelnut relatives and found elderberries, oaks, maples, basswoods, hickory, cherries, mulberries, raspberries, and more—in short, a full array of native species to build a jungle-like environment.

When we placed chickens back in their habitat, their behaviors aligned. Chickens hide from aerial predators amid trees, brush, and ground forages. These perennials cool the ground, increasing humidity, creating conditions for chickens to thrive even on hot days. Chickens scratch and break up capillary water channels, which in turn increase humidity at ground level, which in turn helps grains sprout and forages grow. And chickens provide high-quality fertilizer: Chicken manure contains the thirteen critical nutrients plants need to thrive. On sunny days chickens prefer to be out under the trees instead of in the barn. But once the sun sets, they know to run into the coop for nighttime shelter.

Here on the farm, energy is transformed from thin air, soil, and sunlight into grasses, grains, and vegetables—which, when eaten by animals like chickens, give us eggs and meat. We see ourselves as stewards of these energy transformation processes. When I decided to get into the chicken business, beyond just my own farm, this became the scientific grounding of our system design.

We call our chicken Tree-Range® Chicken and the system Poultry-Centered Regenerative Agroforestry. We work with poultry to reignite naturally occurring biophysical and chemical processes on the land. The people power behind this way of farming comes from farmers wanting to be part of building new systems, consumers seeking nutritionally dense foods, and communities seeking self-reliance, food security, and sovereignty. In the end, we are achieving an all-relatives connectivity, from the tiny microbes in the ground to the tallest tree and everyone in between.

But the human connection and realization of the magnificent design we inherited from the planet itself is the first step. This we call the indigenization of the mind: our realization that to build regenerative systems we must reclaim agency in science, technology, methodology, the land itself, and the supply chain—so that this natural wealth is managed for the purpose of feeding the world and restoring balance to the planetary ecosystems. When we do that, we won't just have chicken, we will have system-level decolonization.

The Cactus Forest

Gary Paul Nabhan

I am winding my way through all manner of thorny shrubs and spiny cacti to revisit some desert farmers I first met more than forty years ago. I didn't quite recognize at that time how much they had to teach about regenerative farming, and now I feel humbled that it has taken so long to appreciate their significance for cultivating a viable pathway to food security in a hot, dry world.

Whenever I feel unsettled about the perilous state of food production on what I call Planet Desert, I remember these former neighbors of mine, who keep up the oldest known traditions of cultivation in what is now the United States. I drive two hours west from my present home over to my former residence in the Giant Cactus–Ironwood Forest. Here, just north of the Mexican border, I continue to learn from these steadfast farmers about how to persevere with agricultural practices that enrich rather than degrade the landscapes from which they draw their harvests.

Around 1984, I first became acquainted with these tenacious descendants of the first farmers in North America, a community that still

31

farms along at least eleven desert washes in southern Arizona. They are the northernmost farming community of *Atta mexicana*—Mexican leafcutter ants, as they are known in English—a species that has gone about its agricultural tasks uninterrupted for at least thirty million years.

Once I've arrived, I begin to see ants bustling across the landscape of their mounds, carrying leaves, flowers, and other plant matter over their heads and backs. Their well-defined leafcutter superhighways provide me with clues to where they have excavated entry points into deep belowground tunnels and growth chambers. In the depths of these chambers, the ants cultivate a food source that looks sort of like kohlrabi but is actually a fungus.

Somewhere in Mexico about thirty million years ago, leafcutter ants entered into small-scale horticulture through an evolving mutualistic relationship with the wild progenitor of a fungus now called *Leucoagaricus gongylophorus*. Each ant colony diligently harvests up to a quarter million separate flowers, buds, seeds, stems, and leaves from the perennial plants in their neighborhood and brings them back as food for the fungi. The fungi enzymatically digest the plant material, then grow special hyphal organs (gongylidia) for the ants to eat.

Generation after generation, these worker ants have maintained an agrarian tradition older than humanity itself, without appreciably depleting any of the resources they rely on over the long haul. The food they harvest for their fungal mutualists consists largely of ten woody perennials, six of which are nitrogen-fixing legumes. Grains from only two annuals—a psyllium-producing plantain and an introduced Mediterranean kelch-grass—make up an insignificant portion of their harvest, less than 5 percent of the organic material processed by the fungi in most years.

By about twenty million years ago, the ants and the fungi were full-tilt into symbiotic relationships, the likes of which would make most human farmers in North America look like isolationists. The *L.*

gongylophorus fungi is now an obligate symbiont of Mexican leafcutters, so dependent on its ant partners that it has never been found "in the wild" outside the agrarian ant colonies. As for *Atta mexicana*, the ants known in Spanish as *mochomos* (night-workers) or *chicatanas* (mule train drivers), they have lost their capacity to make asparagine, a key amino acid for nourishing other animals, because the fungi provide all they need.

It remains an awe-inspiring experience to visit these old-time farmers. The leafcutter ants consume a whopping 17 to 20 percent of all leaf litter in arid subtropical ecosystems, foraging out from their nests at least 130 meters in every direction in harvesting areas of 5 hectares or more. Each colony collects about two hundred kilos of leaves, bracts, buds, and blossoms per year, not bad for ants that average less than 12 millimeters in length. They harvest a diverse desert bounty, including the pods of mesquite, acacias, Palo Verdes, and ironwoods that Indigenous desert dwellers have eaten for millennia, but the primary feedstock they offer their fungal partners is from an ancient medicinal plant, the creosote bush, a drought-adapted shrub with circular clones that may live upwards of 8,000 years.

What seem to be the hallmarks of this profoundly successful agricultural society?

Females are in charge, with the queen of the leafcutters distinguishing herself as the largest ant known in North America. The workers include harvesters and soldiers, which protect the supply chain as harvests come into the underground nests but strike only in defense, never offense. The nests are moved around within 5-hectare areas along ephemeral and intermittent watercourses, so that the harvesting area shifts on a decadal scale, almost like a rest rotation cycle. And the food chain is based almost exclusively on perennial plants that need not be ripped out of the ground and replaced on an annual basis, thereby maintaining stability in microbial communities.

One hundred fifty kilometers east of this agricultural Eden is another ancient site that should command the attention of anyone truly interested in a durable agriculture adapted to the conditions of this continent. At the base of twin volcanic ridges—Tumamoc Hill and A Mountain—Tucson, Arizona harbors the oldest continuously farmed landscape in the United States that was cultivated by *Homo sapiens*. On the floodplain of the Rio Santa Cruz, archaeologists identified chapalote flint corn in a hearth near an ancient agricultural site, radiocarbon dated as 5,500 years old. On historic Tohono O'odham land, but possibly from a prehistoric culture that the O'odham may have displaced, annual cropping systems may have had their earliest arrival north of the Tropic of Cancer in arid zones. Within 3,500 years, large river diversion irrigation systems came into the same watershed and lasted in their prehistoric form until about 1400 BCE.

But when I stand on the Rio Santa Cruz floodplain near the site where those maize kernels were excavated, I always look up at the volcanic ridges above them. In good light, you can see the curvilinear rows of dry-lain stone terraces that were probably used not for maize but for agaves, perennial succulents that produce the same amount of edible and drinkable biomass as maize on a decadal scale but use about half as much water. There are roughly 120,000 hectares of prehistoric agave cultivation mapped by archaeologists in and around the Tucson basin, and all of it was rain-fed, without the diversion of rivers or pumping of aquifers.

Along with ethnobotanists Wendy Hodgson, Vorsila Bohrer, Suzanne Fish, and Paul Fish, I have found on terraces near New River, Arizona, several remnant perennial clones of *Agave murpheyi*—which, like *L. gongylophorus*, is not known to science outside of cultivation, except on abandoned prehistoric terraces. Clearly, people here were cultivating agave at least as early as they were planting corn, well before corn spread into the rest of the present-day United States.

An agave, as you may know, looks like a New World analog of an aloe vera plant, and it is a thorny, succulent perennial still used for weaving into rope and handbags, roasting into a foodstuff, fermenting into a probiotic beverage, or distilling into mezcals such as the now-infamous tequila used in cheap, sugary margaritas worldwide. But unlike most margarita aficionados, agaves have learned how to drink responsibly, using only a fifth to half the amount of water that maize or sorghum uses to yield the same amount of edible or drinkable biomass.

What's more, the agaves that have been in cultivation in arid regions for over 8,000 years have never been just plants; they are holobionts, or communities of many interdependent creatures. Beneficial microbes live on the plants' sword-shaped leaves and in the roots. Certain bat and hummingbird pollinators form strong allegiances to particular stands of agaves. And you can't begin to understand the agave holobiont without accounting for the Indigenous stewards who treat these plants as the sacred incarnation of the Aztec goddess Mayahuel, who ensures fertility and abundance to all involved in her spiritual traditions. Much as a stand of old-growth redwoods contains so much more than just a bunch of individual trees, these agaves help to form a vibrant forest.

This Cactus Forest includes other desert plants too. Although agaves may stand alone in the tequila monocultures of Jalisco's blue deserts today, Indigenous people in this arid part of the world have historically consumed more than 100 native plants in addition to eight species of agave in their Sonoran Desert homelands. Agaves—along with thirty-three other species of desert-adapted succulents that evolved to conserve water by photosynthesizing at night—anchor the soil in many of these Indigenous foodscapes.

These long-lived plants have ensured a consistent supply of nutritious carbohydrates through the harshest seasons, using water so efficiently that they could ride out extreme heat and drought that would kill other species. Mesquite, a leguminous tree with protein-rich pods, has leveraged another strategy for desert resilience. This irrepressible tree plunges its roots far underground, as much as 165 feet below the soil surface, to access deep water. Wild chiles and Mexican oregano, both shrubs, have added antioxidant-rich flavor to desert dwellers' meals. All these hardy perennial plants have also provided a shady nursery, so that when rains finally come, less heat-tolerant annuals like tepary beans and chia can find life in their shadow. Together, I call this agrarian collective a desert nurse plant guild.

Again, that term tends to focus on the plants rather than the entire holobiont superorganism of plants, bacteria, mycorrhizal fungi, and other microbes that function together to determine leaf shape, color, size, flowering time, disease resistance, stress tolerance, and many other traits we find in these perennial crops. They rely on what we scientists call "epigenetic influences on their phenotypic plasticity," which is just a nerdy, geeky means of affirming that the whole is greater than the sum of the parts. While both Mexican leafcutter ant agriculture and agave agriculture have perennial plants as key components, they would not succeed without coevolved microbial communities that allow them to thrive under the hot and dry conditions of the Sonoran Desert.

Our present climate predicament has inspired me to invite many proponents of perennial agriculture for pilgrimages in this desert, so they can see that it is a laboratory for the future of agriculture across two thirds of North America and much of the food-producing terrain on Planet Desert. According to a 2017 analysis by the United Nations Food and Agriculture Organization (FAO), at least 44 percent of the world's food is produced in arid and semiarid lands, now with increasingly high water and energy costs and rising death rates from heat and dehydration among farmworkers. If the FAO predictions are correct, four fifths of the planet's surface currently planted in crops and three fifths of its food yields could be disrupted or permanently damaged by climate change. And yet with few exceptions, the current breeding goals for annual crops do not factor in 110°F to 115°F temperatures during flowering and fruiting, the loss of a third of the river water formerly used in the US West for crop irrigation, highly saline soils, or even highly damaging ultraviolet radiation.

Decades ago, I myself sang the praises of many annual desert food plants, but my own agroforestry efforts at home have increasingly turned to the agave, the perennial chiltepín, the deep-rooted mesquite, and the water-thrifty prickly pear. And yet the genius of ancient desert food systems lies not with any one "miracle crop" but in the way they work together. We must heed the lessons of the Indigenous agave stewards, and the leafcutter ants who came before them, and join in their enduring effort of co-adaptation.

Our future food systems will look, taste, and function more like diverse Giant Cactus–Ironwood Forests than the dystopic monocultures that stretch across so much of rural America. Let us not wait for tens of thousands more acres to go belly-up as rivers cease to flow and aquifers fail to percolate up before we take the desert's wisdom seriously. As Joseph Wood Krutch once summed it up so elegantly, "In the desert . . . the very fauna and flora proclaim that . . . one kind of scarcity is compatible with, perhaps even a necessary condition of, another kind of plenty."

A Food Forest for Southeast Atlanta

Kelsi E. Bowens with Rosemary W. Griffin

Summers in North Carolina with my grandparents and aunts planted the seeds of my love for fresh food and the environment. I remember the dark soil and patchy green grass under the pecan tree in the backyard. Although we came over to play, we had to work too. After helping my grandmother in the garden and hanging washed clothes on the line to dry, I would treat myself to a cool reprieve from the summer heat beneath the canopy of pecan and southern magnolia trees.

My grandmother, who had grown up on a farm, took pride in her gardening skills, and it showed. As you walked up to their red brick house, you were greeted by a flower patch bursting with purple and multicolored petunias, roses, and a large snowball bush—one my brother and I often crashed into while trying to slow down our bikes. My grandfather, who also grew up on a farm, kept a row of tomatoes in the front yard. As a kid, I didn't care much for tomatoes, but I grew to love them, especially when my grandmother fried the green ones or sliced the red ones for tomato sandwiches slathered in pepper and mayo.

In the backyard, they had a chicken coop, hunting dogs, and a garden full of mustard greens, collards, turnip greens, okra, and squash. There were even apple, fig, and peach trees along with the pecans! My grandparents showed me that growing your own food wasn't just a rural tradition—it was a skill passed down through generations and something you could bring with you wherever you called home. It was also an act of defiance: My grandmother and a neighbor planted a large flower bed on a neglected city-owned property on the corner of their street, fighting back against the stereotype that Black people don't take care of our neighborhoods.

Whenever I returned to Atlanta, I questioned why the abundant fresh food lifestyle of my grandparents' house couldn't be found in our community. My great granddaddy JD lived on Winton Terrace in Fourth Ward, a historically Black inner-city neighborhood about 1 mile from downtown that has become unrecognizable due to gentrification over the last two decades. His backyard was filled with muscadines, cabbage, peppers, and collard greens that we would pick fresh for holiday dinners. But in my own neighborhood, these edible landscapes were missing.

I grew up in southwest Atlanta, another inner-city neighborhood close to downtown. A notable characteristic of Atlanta—and one that I know firsthand—is the presence of a thriving African American population and middle class. And yet, as a young Black girl with a passion for the environment, I was keenly aware that many people in my community have been intentionally disconnected from fresh food systems. Despite agriculture being the largest industry in Georgia, one in eight people are considered food insecure in our state. When you zoom in on Black populations in Fulton County, that number jumps to nearly one in four. Even in a city that has had Black leadership since

1974, the lingering legacy of racism and inequality is clear, particularly in the environmental and social vulnerabilities that still disproportionately affect us.

So at age twenty-six, I found a chance to turn my passion into action. I landed my first conservation job and had the opportunity to collaborate with residents from Lakewood and Browns Mill, two predominantly Black communities on the southeast side of Atlanta. Alongside government partners and nonprofits, we created what would become the largest food forest in the country.

Back in 2016, the US Department of Agriculture Forest Service granted the City of Atlanta seed money to purchase 7.1 acres that would become the Urban Food Forest at Browns Mill. That seed money was leveraged to create a community-driven concept plan, including a workforce development program for young adults from the area to build a community garden and trails throughout the property. Many local residents and partners were part of creating this eco-educational space that I've grown to love, including Rosemary Griffin and Doug Hardeman. Friends since they met in kindergarten more than sixty years ago, these two dedicated volunteers proved indispensable to the fledgling garden.

I first met Rosemary and Doug at the Food Forest for a fall festival. Doug wore a navy blue shirt with the sleeves cut off, khaki shorts, and glasses. He introduced himself and asked, "So who's taking care of all of this now that y'all built it?," nodding his head toward the fresh cedar raised beds and trellises that came from the workforce training program. He must've asked in jest because the plants looked pitiful. And honestly, we didn't have a great plan at the time for how to recruit more volunteers and community members to maintain and adopt the beds. "It's that bad, huh?," I responded. We all laughed, and I let them know we were looking for help. Doug smirked at Rosemary and looked back at me. "Good luck with that."

But within a week, the two showed back up. Doug drove a burgundy flatbed while Rosemary had a sporty red Mustang. This time, they came ready to revamp the area and establish a thriving system for planting. Rosemary sported a straw hat and gardening boots and was full of quips to keep the day rolling as they worked in the blistering sun. She made it plain that she wasn't a "dig in the dirt kind of girl" and that "Doug dragged me out here." But time proved otherwise. They were both inquisitive and consistent, showing up every day to steward this space and grounding us in the reality that it would be years and a lot of labor before this seven-layer food forest would solve the food equity issues that plague our city. Doug took on the role of community garden manager, and Rosemary became the assistant garden manager. Over time, alongside a growing group of volunteers including Celeste, Elizabeth, Harold, Judi, Mike, and Maurice, they breathed life into the community garden that we see today.

Before we could afford an irrigation system, they wielded buckets of water that kept the wither at bay. Before the hoop house, they nursed the seedlings to life. Within a matter of months, over thirty garden beds blossomed with peppers, eggplants, cucumbers, parsnips, and more. Perennial vines of muscadines and raspberries intertwined along the trellises that had once stood barren.

As much as the volunteers poured into the place, it returned the favor with other benefits of gardening. Benefits like burning calories, increasing vitamin D, relieving stress, lowering blood pressure from the work you put into the healthy harvest you reap. Most of all, the food forest continues to provide a source of community. Although the yield from this young forest has yet to solve larger-scale food access issues, it offers learning opportunities in every layer, and the garden is just one piece of this purpose-built land.

Over a hundred pawpaw, American persimmon, apple, and nectarine trees joined the ranks of the stately pecan and red mulberry trees planted in the 1970s by Mr. Willie and Mrs. Ruby Morgan, a couple who once owned the land and operated a farm for many years. Community members still share stories about how the Morgans would hang grocery bags of excess harvest along the fence for their neighbors to enjoy. I'd like to think that Rosemary, Doug, and many other dedicated volunteers carry forward the agrarian Southern hospitality of the land. The pawpaw, persimmon, and mulberry trees not only produce fruit year after year but also provide habitats for birds and pollinators while improving the soil and microbiomes in the land.

Trails lined with inoculated mushrooms lead us past a stack of humming honey beehives, bat boxes, birdwatching opportunities, and a hoop house that's home to cooking classes led by Doug's daughter, Shandrea. Khari's hexagonal compost bins ensure that nothing is wasted. Celeste's herbal and medicinal gardens—filled with rosemary and lavender—provide fresh herbs to make teas. These woody perennials are low maintenance and sequester carbon as natural solutions to the urban heat island effect here in Atlanta. On the back end of the property, a small tributary of the South River meanders alongside a medicinal garden. And Soisette, an impassioned leader of the Friends of the Food Forest, advocates for continued programming at the forest.

This forest grows because of the persistence of its people, showing up day after day to steward this space. Unfortunately, Doug passed away in 2021, but Rosemary, Doug's wife, Yvette, and daughter Shandrea still carry on his legacy. Today, most of the harvest from the food forest goes to local church pantries, regular volunteers, and residents around the food forest. It serves as a model for urban areas worldwide and showcases how perennial systems can restore ecosystems, educate, and provide solutions for food insecurity. Doug's legacy, my grandmother's legacy, and this Southern land keep me grounded in this work. But there's always more to be done.

How many other communities have the same story? Where assets come, but the brunt of the work remains? The reality is, building and maintaining a food forest takes a ton of commitment and resources to continue to benefit the people it was modeled to support. The community where a food forest was built to address food access inequities certainly can't be the sole community that stewards the forest too. And yet most days, Rosemary shows up—now alongside Yvette—to ensure that seedlings are sown, tended, and harvested, again and again.

To the world, this space might be just a food forest. But to me it's an outdoor sanctuary that nurtures our innate connection to the land. It's my sincere hope that if you're reading this, you'll show up and support this community. As you marvel at the growth within this young regenerative space, bring a friend, grab some gloves and as my grandmother would tell me as a kid, "Get to work." Let them know that we sent you.

~

In honor of Doug Hardeman and Mae Belle "Inma" Eccles; special thanks to Trees Atlanta, AgLanta, The Conservation Fund, the USDA Forest Service, and all the other organizations that played and continue to play a role in sowing the seeds of this place.

Neighbors

Megan Kaminski

We speak in the language of elderberries
growing in back alleys, boneset salves
and tulsi tinctures passed around as
needs call. You bring a bowl of Asian pears
to repay me for mulberries this spring,
you claim, but we never keep track.
We share a currency of green life; despite
lean times for us teachers, love gives freely
in soil and the abundance we nurture daily.
The woman building next door had all the
trees cut down, a nuisance to her eyes.
When you were planting okra and sage
she offered you money to "landscape"
her yard next spring, seeing only the color
of your skin, your knees on dirt. She doesn't
understand this practice of tending those who
give breath and compost grief into hope,

that our labor yields more than kale and scenery.
New houses keep getting bigger, crowding
us and the pawpaws out. She's building
a stone wall against my bedroom window;
"I worked hard for this," she repeats.
And you shake your head as we toss
sliced peaches and sun sugars with basil,
pour a few glugs of olive oil on bread to savor
with the backyard calling cicadas and kids.
Swatting mosquitoes off bare limbs, we
inhale all we can of this late August day.

From the Fringe

Keefe Keeley

A few years after my grandparents died, my mom and her siblings sold the family farm in Iowa to a cousin who ran a bigger farm. My mom and my aunt used the proceeds of the sale to buy a patch of gorgeous but somewhat worn-out ridgetop farmland, about 5 miles from where I grew up in the hilly Driftless Area of Wisconsin. I'd been recently convinced by Wendell Berry's *Unsettling of America* that all sorts of societal ills spring from young folks pulling up roots and pursuing highfalutin city careers while the rural communities they come from wither on the vine. So with all the idealism and accumulated wisdom of a twenty-two-year-old and enough privilege to go off the beaten path, I decided to try my hand at farming up on that ridge.

I had no idea what I was doing. To learn, I got a job on an organic farm in the valley down the road. We grew vegetables in the rich soil that had eroded down off ridgetops in the 150 years or so since settlers started plowing. Among the veggies on the farm were a handful of fruit crops: strawberries, raspberries, currants, rhubarb. When these perennials ripened into enough abundance, these were the crown jewels of that week's produce boxes.

I began planting my own perennial crops up on the ridgetop, hoping to hold onto the topsoil we had left. Masanobu Fukuoka's *One Straw Revolution* had introduced me to biomimicry (i.e., farming the way nature works), and Mark Shepard's *Restoration Agriculture* had inspired me to plant polycultures of chestnuts and hazelnuts, as well as hickories, walnuts, apples, plums, hazels, elders, brambles, currants, and grapes. One day, a dozen or so friends joined to plant trees in widely spaced rows in one of the hay fields. This planting design—alley cropping—allowed the farm to continue to produce hay, intercropped in the alley between the tree rows, while the trees matured.

Those trees continue to grow, but I still struggled to make a living on that farm. On other farms, some folks have figured out how to make a go of intercropping perennial crops with other farming enterprises. I figured they were surely cleverer, more focused, and better at building fences to keep deer away from trees than I was.

Eventually, I came to peace with the reality that my gifts are not best suited to being a farmer, but I also gained perspective that helped me see it wasn't just my shortcomings that kept a perennial farm dream out of reach. As I pivoted out of farming, I went to work at Wisconsin's Department of Agriculture and then to graduate school. A bigger picture came into focus as I worked with other land stewards and studied the state of our food system. I saw that even the most talented, hard-working, and well-off farmers had a hard time breaking the mold of annual monocultures and shifting to more perennial polycultures. They were locked in by numerous obstacles: A handful of crops receive the lion's share of R&D, subsidies, and insurance; competitive advantages are conferred by externalizing costs; and certain consolidated market players favor a race-to-the-bottom system where maximum production of commodities ensures bottom-dollar prices.

But I'd seen alternatives take root and grow, despite the obstacles. My dad worked for thirty years at Organic Valley, a farmer-owned market cooperative that my father-in-law, who helped found the co-op, likes to call a social experiment disguised as a business. I had the benefit of seeing this experiment succeed and provide farmers with opportunities to break the mold. When I was a kid, Organic Valley was a shoestring operation in what was widely seen as a fringe movement for luddites and hippies. Now it is a top brand in an organic food market representing 6 percent of all food sales in the United States. There's still plenty of room to grow, but organic farming has moved from the fringe to the mainstream and represents a viable option for an increasing number of farmers.

~

Despite my best efforts to put my head down and focus on grad school, the perennial fringe beckoned. I got plugged in with a group of farmers, graduate students, and community members in central Illinois, where King Corn's empire is nearly uninterrupted, not even by Queen Soybean. Yet here was this upstart group, unbowed and digging into biomimicry, doing participatory research on woody perennial polycultures. They were calling themselves the Savanna Institute. Right up my alley crop.

In my research, I'd been obsessing over savannas: fascinating ecosystems where scattered trees dot expansive grasslands. My experiments in cooperation with farmers were testing how prescribed grazing by their cattle could mimic extinct megafauna and help restore these ecosystems on their farms. Learning more about savannas was challenging my understanding of biomimicry; was I being too simplistic saying nature had the answers?

In my background reading, I learned that a slow-motion epiphany in the 1980s and 1990s uncovered a Midwestern plant community that wasn't just a mix of two ends of the spectrum between forest and prairie. Practitioners in the emerging field of restoration ecology described distinct assemblages of species adapted to living among scattered oaks, hickories, and other woody plants. This plant community featured more shrubs, more soft-bodied fruits, more broad leaves adapted to partial shade. Many of these species began to show their face only when people restored a disturbance regime: removing invasive shrubs, thinning the tree canopy, and returning fire to the land.

Disturbance regime sounds like the name of a punk rock band, but it's even cooler than that. *Disturbance* refers to an event that turns back the ecological clock—like a fire, a scouring flood, or a grazing herbivore—and *regime* is the frequency and intensity of disturbances. Savannas are poster children for what's known as the intermediate disturbance hypothesis. These ecosystems require enough disturbance to prevent trees from growing into a closed-canopy forest but not so much that all the trees disappear.

Of course, while the hands-on approach of restoration ecology helped "rediscover" the Midwestern oak savanna, it was lost from the landscape and common knowledge because of the displacement and attempted erasure of Indigenous people who were and are partners in this ecosystem. Tens of thousands of years ago, before humans were here, mastodons and giant beaver and their ilk were doing the heavy lifting on reducing tree cover between intermittent natural fires, but at some point, humans took the baton and became primary agents of the disturbance regime.

I can't pretend to know much about the whys and the hows of Indigenous land care of oak savannas, although I've been grateful to learn when I can. At the broadest level, what is most important is to recognize that humans are a key part of this ecosystem. I'd been thinking biomimicry was about learning from something in nature that humans have nothing to do with, like using the shape of whale fins to design wind turbine blades. But oak savannas aren't just some random cool biological phenomenon we can riff on. They are an *agro*ecosystem that developed over millennia, with humans a part of the biology, as a keystone species, in fact.

Restoration ecology is one way of reestablishing people's key role in these ecosystems. The discipline is beginning to learn from traditional ecological knowledge while avoiding unjust appropriation of that culturally situated knowledge. And yet the cost and intensity of effort required for nonremunerative restoration projects typically limit this work to small, isolated pockets of land, which are mere postage stamps on the larger envelope of landscapes. It takes a lot of skillful disturbance: soil prep, planting, weed control, mowing, harvesting seed, adapting to weather, and more. Guess who besides restoration ecologists are really good at doing all this land management? Farmers! And not only are farmers good at those activities, they have land! If we're going to meet ambitious goals for water quality, wildlife connectivity, and climate change mitigation, we need restoration across the larger landscapes that farmers steward.

~

As I witnessed grazing animals restoring disturbance regimes to Wisconsin and polycultures of perennial crops supplanting King Corn in Illinois, I saw what savanna biomimicry was all about: Rather than taking land *out* of agriculture in order to restore its ecological health, we needed to restore ecology *via* agriculture. True to the deep history of humans in savannas, farmers were the keystone species to make that happen. If we could begin to change the equation, to open a viable niche for the perennial farmer, here in the heavily industrialized Midwestern farm landscape, by golly we could do it anywhere. Or, rather, folks in all the other anywheres out there might see that the monarchy of monocultures is not as immutable as it might seem.

Since I began to recognize that the challenges I faced as a beginning farmer sprang not just from my foibles but from systematic barriers to farmers beginning or transitioning to more perennial agriculture, it's been a profound privilege to try to lower those barriers through our work at the Savanna Institute. Working first with founding folks in Illinois and with farmers in my home landscape in Wisconsin, and now with farm and community partners across the Midwest, we have been defining those barriers and building out efforts to systematically dismantle them. Demonstration farms. Improved perennial crop cultivars. Technical assistance for farm design and planning. Research to better understand what perennial agriculture can do for soils, water, wildlife, and climate. Educational materials, training, apprenticeships, and communities of practice to learn farm management strategies. Access to land, capital, and markets appropriate for diverse perennial crops. Opportunities for farmers from underserved communities, especially communities with traditions and values that align with perennial agriculture. These efforts are rooted in a vision of diversity and abundance, where farms provide more than commodity production and contribute to all that we depend on: clean water, biological diversity, a stable climate, healthy soil, buffers against floods. Many of us also look to farms as pillars of vibrant agrarian communities as we search within our neighborhoods for pathways to a more just and prosperous world.

I think of these efforts as part of a broader disturbance regime. As we disturb the status quo of the industrial ag complex, we can make space for what is currently fringe and dormant to become more established. Our work is to be ready for these disturbances, be agents of disturbance where prescribed, and build capabilities to nurture the germinating seeds and resprouts of perennial agriculture to fill in the openings as they appear on the cultural and physical landscape.

In this broader social frame, not every disturbance implies violent or sudden change; it might just be something from the fringe making its way into natural openings over time. Up on the ridgetop farm, the tree crops my friends and I planted continue to grow in the rows stretching along widely spaced contours of the hayfield. Each year the neighbor continues to make hay in the alleys between these tree rows. A couple years ago, I happened to pass that neighbor's home farm, where they raise corn and soybeans across most of our shared ridge. To my delight, plantings of trees and tallgrass prairie strips had begun appearing on the borders—on the fringes, that is—of their fields.

In the Presence of an Olive Tree

Omar Tesdell

What is this crisis we find ourselves in? Is our human society at risk? Many people now ask how we will survive the coming violences and migrations and new forms of human exploitation. These are questions that unfree peoples have always asked. War surrounds me as I write these words.

In Palestine, human cultures and violences have come and gone for many generations. As the Palestinian scholar Edward Said has said, drawing on Antonio Gramsci, these cultures have left an infinity of remnants. However, the remnants have no inventory or guide. So the main task is to make one.

Palestine is the southeastern corner of the Fertile Crescent, a strip of land at the eastern end of the Mediterranean. Not a desert but a Mediterranean climate, much like that of California. The soils were built over millennia by the original vegetation in the hill region of Palestine: the oak (*Quercus*) and terebinth (*Pistacia*) trees that spread across the landscape when the Mediterranean receded and the climate shifted. It is the Holy Land, a small place loaded with meaning for Jewish, Christian, and Muslim faith traditions. It is the ancestral homeland to many major agricultural crops that feed people around the world, such as wheat, barley, oat, pea, lentil, and flax.

The wild relatives of these plants grow in the hills around my home and in the sidewalks. The words that we speak in Palestinian dialect today like بعلي ba'ali or "rain-fed crops" are also like living seeds. Such words predate the coming of Arabic to Palestine and have remained in our culture for more than 1,500 years.

Our people, like so many others around the world, are oppressed. The community teaches me about human freedom and adaptation every day. We live, go to school, do perennial agriculture research, collect plot samples, and go to soccer practice in this context.

～

There is really no way to *know* what will be useful. What we think might be a saving remnant may not be recognized in the future. We cannot force recognition of our work. We can make it capable of being recognized at a future time. *It is a subtle but very important distinction between thinking you know and making it possible to be known.*

Plants remain in and among the hundreds of Palestinian villages destroyed after 1948. In the classic Palestinian film *Ma'loul Celebrates Its Destruction* by Michel Khleifi (1984), internally displaced Palestinians visit their former village. In a particularly moving scene, an older man tries to locate his former house. Among the Aleppo pine trees (*Pinus halepensis*) planted after its destruction, he searches for an olive tree (*Olea europea*), several different prickly pear cacti (*Opuntia ficus-indica*), a carob tree (*Ceratonia siliqua*), a fig tree (*Ficus carica*), an almond tree (*Prunus dulcis*), and a mulberry tree (*Morus*). The remaining trees serve as landmarks after the stone homes themselves have been demolished.

Similarly, the sacred aromatic herbs of each Palestinian kitchen, *maramiyya* (wild Greek sage *Salvia fructicosa*), za'atar (*Origanum syriacum*), za'atar sabbal (*Thymbra spicata*), za'atar rumi (*Satureja thymbra*), za'atar farisi (*Thymbra capitata*), and za'atar balat—among so many others—have grown, flowered, died, and given seed each year since 1948, in and around the villages of Palestine. They remain, recognized or not, in a remnant geography of herbs that were once harvested and may again be used by future generations.

~

Living in the midst of 10,000 years of agriculture, I started thinking about what might be. What could possibly be done when I am in the presence of an olive tree that has been feeding people with its oil for hundreds of generations?

We started with our rare native scrubby oak species, which I like to think of as the "soil works" that builds our land. Reaching under the branches of these low scrubby trees, even in the summer dry months after weeks without rain, one can smell soil in formation as the leaves decompose. Beginning with the Aleppo and Kermes oaks, we learned when and how to collect the acorns while respecting the animals who feed and depend on them. Now we have various species in our collections and have been propagating and outplanting from these. But while planting seeds so often represents a beginning, we only really activate a remnant that someone or some process has left behind.

A fruit is a seed-bearing structure that allows flowering plants to spread their seeds. A root is a more universal structure that both supports vascular plants and transports water and nutrients to them. Both seeds and rootstocks are technologies of time travel. They bend and twist time and place to suit their purposes to increase the possibility of spreading. They embody historical legacy and future potential. It seems that we can only invest our energy in starting processes that may bear fruit in the future. These processes must be built on our understanding of historical reality that brought us to the particular moment. So there are no guarantees.

~

Our research group emerged from conversations between farmers and researchers in Palestine. Our work has been made possible by a team of bright young Palestinian researchers and colleagues around the world. We conserve wild food plants, gather local knowledge about them, and build digital data structures to drive our research.

We are technologically and globally connected but grounded in our community of people and plants and other beings that make our landscape. We share our work in international scientific journals as well as self-published research conducted by people in their own places. We aim to open spaces in which more just and adaptive societies can become possible.

~

Many remnants stem from the oppressions of human societies built on annual grain-based production. We should recognize our responsibility and culpability in the remnants that lie around us. Making right today and making room for future remnants of tomorrow. If a tree is known by its fruits, how might the adaptations we learn become the fruits of tomorrow?

Like the tree, we release the fruits and cannot control what happens next.

Makwa-Miskomin: Bear's Red Berry

Wendy Makoons Geniusz

Chi-mewinzha, in that long-ago time, when humans were new to this world, Gichi-Makwa, the Great Spirit of all the Bears, slid down a sand dune beside a lake. In the path where Gichi-Makwa slid, a little vining plant with leathery leaves and red berries was created, the first Makwa-miskomin. Then Gichi-Makwa popped one of the red berries from the plant into the mouth of a crying human baby. The baby immediately stopped crying and started swallowing repeatedly. "There," Gichi-Makwa said, "now he has his last berry, and he won't have to cry anymore because he can keep down all his other berries." Our elders say that this "last berry" is the little piece of skin that hangs in the back of our throat and that Makwa-miskomin means "Bear's Red Berry." Mii sa iw. Miigwech. That's it. Thank you.

I have known Makwa-miskomin since I was six years old. I remember when I was first introduced to this plant because it was the week I received my Anishinaabe name from an elder Anishinaabe medicine woman, Keewaydinoquay Peschel, who was my mother's teacher. Keewaydinoquay learned these teachings as a child living on Michigan's Leelanau Peninsula in the 1920s and 1930s. The story above is part of a much longer Anishinaabe story about the creation of two botanical beings, Nookomis Giizhik, "My Grandmother Cedar" (also known as white cedar, *Thuja occidentalis*) and Makwa-miskomin (also known as bearberry, *Arctostaphylos uva-ursi*). Whenever we are teaching about these beings, it is important to tell this story.

Makwa-miskomin clings to sandy hills, dunes, and small cliffs around bodies of water. The plant's diminutive leaves are shiny and rather thick, so that they bounce back when bent. Many humans do not see this life form, who lives so close to the ground. Other plants, even blades of grass, tower over Makwa-miskomin and are more noticeable. Yet this plant's vines form an organism that can grow to be massive. A stand of this perennial plant can be a century old, although many are ripped up long before they reach that mature age.

I find sitting beside this plant very calming because Makwa-miskomin is not in a hurry. Unlike some plants, Makwa-miskomin does not have to rush to consume sunlight during a few months of the year because this plant's leaves are attached to the vine and green all year long, albeit sometimes under feet of snow.

The vine of Makwa-miskomin grows very slowly. The "new growth," on the very end of the vine, is a pliable, supple stem. If one follows this green stem back into the plant, soon a thin, stick-like stem appears. This stick-like portion takes a while to form and can be several years old. Further back, often covered by sand or newer vines, is an even thicker, dense vine—possibly decades old, having added a thin woody "ring" each year. But do not go digging for that old growth, or you will hurt this elder.

There is a stand of Makwa-miskomin in Wisconsin that I have been visiting for decades. I go there in mid-August, after the plant has had the summer to grow. Long, green tentacles reach off the small sandy bluff beside the lake, where this plant lives. The leaves on the top of the dune grow in smaller clumps, reaching up to the sun. The smooth leaves, split with a single line down the center, shine dark green and curl down around the edges, away from the sun. If it is a good, wet year, Makwa-miskomin has longer green stems. One time they were a foot long, giving us lots of medicine.

I approach this stand of Makwa-miskomin by walking from the beach toward the sandy hill because it is easier to see the vines from the beach than it is from the top of the dune. I bring a small offering of Kinnikinnick, a mixture of local herbs my Mom, Mary Siisip Geniusz, and Keewaydinoquay taught me to bring to elders whenever I want to ask them something. I also, as they taught me, greet Makwa-miskomin by name, and then I introduce myself by my clan, my home, and my name. Then I ask this old being if I may take some of the new growth to make medicines for myself and my family, promising not to injure the plant or to take so much that this plant relative cannot continue to thrive in this spot. These are all Anishinaabe teachings that I learned as a child. If there is no objection, such as a sudden downpour, windstorm, or bee sting, I stand on the beach and clip the green stems right off the dune. Sometimes I stand on one of the many driftwood logs, so I can reach even higher up onto the hill.

I have been taught to use only the new, green-stemmed growth of Makwa-miskomin and to harvest this plant by carefully lifting up the vine and snipping it with sharp gardening clippers above the woody part of the stem, so as not to interfere with the older part of the plant.

I gently put what I have gathered into a paper shopping bag by my feet. My mother always said it was important that the plants I gather go directly into a bag so that they do not come into contact with spores that might be on the ground. The paper bag will allow the plant to dry, unlike a plastic bag, which will cause the leaves to mold. Labeling the bag is important, or one may end up with a series of bags containing unidentifiable plant material.

There are a few other things to know. First, my mother and Keewaydinoquay were both ardent believers in modern medicine. Both women, when they had access to healthcare, went to regular doctor appointments and filled prescriptions at the pharmacy. My mother always encouraged me to get diagnosed by a doctor, especially before taking any of my own medicines, just to ensure that I was treating the right problem. So anyone reading this who has access to healthcare should certainly try that route first.

Second, when in the field, make sure to correctly identify Makwa-miskomin because there are a few plants that look like this one. One of the easiest ways to do this is to find a field guide of woodland plants and look up Makwa-miskomin by the non-Indigenous scientific name, *Arctostaphylos uva-ursi*.

Finally, it is also important to remember that generations of Anishinaabeg have worked to learn, perfect, and preserve this knowledge for our future generations. We as Indigenous people need to be acknowledged as an important part of the history and continuation of these teachings. The general public's increased interest in native plants and local, wild foods has been hurting the Native peoples from whom the knowledge of these plants and foods originated. In many cases, those taking "wild resources" do not leave enough for the people who have been gathering these plants for millennia. Anyone choosing to gather some Makwa-miskomin needs to remember that the Anishinaabeg are still here and still gathering our ancestral medicine.

Now, having given readers all those important disclaimers, here are some of the ways I was taught to work with Makwa-miskomin. According to Keewaydinoquay's teachings, which she passed on to my mother, Makwa-miskomin and Nookomis-giizhik were created at the same time to give humans a way to connect with the other beings of this world and ask for help and guidance. These two plants are the two essential ingredients in the Kinnikinnick we make as an offering to the Manidoog, Spirits, and any human or other-than-human elders—like the plants, trees, and animals—from whom we need assistance. I do not pick Makwa-miskomin specifically for this purpose, however. I use the previous year's stash of Makwa-miskomin in my Kinnikinnick, and I keep the newest harvest for making medicines to take internally. It is important that lives not be wasted, especially when gathered with an offering, so I am not discarding this plant's gifts, simply applying them to another purpose.

There are many medicines within Makwa-miskomin, and this plant is important to other cultures too. My mother taught me to use this plant whenever I have my period or whenever my body is retaining too much water, because a cup of tea made from this plant will make a human body pee out retained water. To make a tea from this plant, I take a sprig or two of the dried plant, pull the leaves off of the stem, put approximately 1 teaspoon of them in a mug, or, if I have one, a tea strainer. Then I boil water in a kettle, and when it boils, I take it off the heat and let the water sit in the kettle for five minutes, so as not to scald the leaves. After five minutes, I pour the boiled water into the cup, where I let everything steep for four to five minutes. If possible, I strain out the leaves before I drink this tea because it gets stronger the longer the leaves sit in the hot water. Sometimes I sweeten the tea with honey. Being aware that this is an important medicine, I do not throw out the leaves, but instead I put them outside on the earth, or I dry them on a paper towel and add them, once fully dry, to my Kinnikinnick.

The berry, after which Makwa-miskomin is named, has to be eaten very carefully because there is a large seed in the middle of it, and it dries out one's mouth. Very few berries are on a plant, and we do not eat them as we eat other berries. The tea, made from the leaves, has a distinct yet mild flavor. Makwa-miskominaaboo, Bear Berry Tea, tastes like a plant that spends as much time as possible soaking up the sun and slowly releasing the essence of that energy through thick, green leaves.

For anyone wishing to work with Makwa-miskomin, please remember that this is an elder who needs to be treated with reverence. Although the vines look little and are often overshadowed by taller plants, a stand could be much older than the human gatherer. Our ancient Anishinaabe story about this plant, a brief version of which begins this essay, is so important that we consider the story itself to be a living, cognizant being who knows when they are being told. This story tells us that people were not fully human when the first Makwa-miskomin was created. We did not have a crucial part of our anatomy, the uvula, until this medicine plant came to us. Makwa-miskomin has an ancient history with humans and a very old spirit. So be courteous when approaching. Introduce yourself and ask for this being's assistance. Then, when you pick some of the stems, make sure to not yank at them because this will uproot the older parts of the plant. Make sure that you only clip the green stem parts. We are in a changing world, a world in which we are going to need all our elder plants and trees if we expect to survive. So we need to treat them with respect.

The Odyssey of Coffee

Ivette Perfecto

It all began with an unassuming perennial shrub nestled under the lush canopy of northeast Africa's forests. This shrub, bearing vibrant red fruits, induces exhilaration and a feeling of excitement and well-being in those who consume it. It grows in the understory of the mountain cloud forest in Ethiopia's Kafa region, one of the few places where it still thrives in the wild. There, seeds are dispersed by monkeys and birds, like the black and white colobus monkey and the silvery-cheeked hornbill. This natural system has been used and maintained for centuries by local communities.

Many tales recount the discovery of coffee's stimulating properties, yet the most beloved involves an Ethiopian goat herder. According to legend, he observed his goats' heightened energy and enthusiasm after they consumed berries from the coffee bush. Whether fable or fact, the geographic substance of coffee's beguiling beginnings remains undisputed, and at least by the late 500s CE the Oromo of Ethiopia were consuming coffee regularly. From there, the remarkable beverage made its way to the Arabian Peninsula. By the fifteenth century, coffee cultivation was flourishing in Yemen and trading hubs throughout the region, later extending to Egypt, Persia, Syria, and Turkey. Coffeehouses burgeoned across the Near East, evolving into significant venues for the exchange of information, ideas, and culture. And sometimes deep political ideas were kindled by this stimulant.

By the seventeenth century, travelers had introduced coffee to Europe, where it initially faced intense opposition from religious zealots, who branded it the "drink of the devil." Nonetheless, coffee eventually won over even Pope Clement VIII. After tasting it, the pope reputedly declared, "This Satan's drink is so delicious that it would be a pity to let the infidels have exclusive use of it." Coffeehouses proliferated throughout Europe, becoming hubs of intellectual cross-pollination where Enlightenment ideas, scientific discoveries, and political discourse thrived, much like their Eastern counterparts over a century earlier.

Ironically, much of the coffee fueling European intellectuals in their debates over freedom and democracy was sourced from colonies in the Global South. The Dutch East India Company, which ascended to become the world's wealthiest company—often considered the first multinational corporation—stole coffee seeds from Yemen, establishing plantations in Java (Indonesia), Ceylon (today's Sri Lanka), and Malabar (India), thereby monopolizing the coffee trade in the East Indies. The French and English soon followed, establishing their own colonial plantations. The working conditions on these South Asian coffee plantations were grueling, marked by meager wages, poor living environments, and rigorous control by European planters—effectively a system of forced labor, often involving indentured servitude or debt bondage.

Although historical documentation of these coffee plantations is minimal, existing paintings and photographs suggest that they differed starkly from the diverse native coffee forests of the Ethiopian highlands.

Sparsely planted, with little genetic diversity, these monocultures of coffee were ripe for a pest outbreak. When the *Hemileia vastatrix* fungus arrived on the scene in the late 1860s, causing coffee leaf rust, it devastated plantations in Ceylon and South Asia. Most were abandoned or replaced with tea, and today we are familiar with Ceylon tea rather than Ceylon coffee.

The demise of coffee production in South Asia led to rapid growth of coffee farms in the Americas, and the geography of these operations differed in some important ways. Whereas coffee farms in Brazil were established primarily in flat areas, like their South Asian predecessors, coffee producers in northern South America, Mesoamerica, and the Caribbean established their farms in mountainous regions and needed trees not only to preserve the health of the coffee bushes themselves but also to prevent soil erosion. By this time, slavery had been largely abolished, but the plantation system continued with a new source of cheap labor: exploited migrant workers who faced slave-like conditions and often debt bondage. And yet some of these workers learned how to grow coffee and started cultivating it themselves. After the harvest season, they would go back to their villages with coffee seeds and incorporate them into their traditional agroforestry systems.

Eventually, coffee leaf rust made its way to the Americas, first recorded in Brazil in 1970. Widespread panic ensued among coffee farmers, who feared a repeat of the devastation experienced in Ceylon a century earlier. In response, the agricultural community hastily recommended fungicide applications and the removal of shade trees from coffee farms. Despite a lack of supportive data, the prevailing assumption was that shade trees fostered humid conditions favorable to coffee rust, an idea later disproven. Simultaneously, the US Agency for International Development initiated programs to enhance coffee productivity throughout Central America, aiming to help countries bolster their exports and address mounting international debts accrued in the previous decade. These programs, coupled with the rust scare, propelled the "technification" of coffee, involving the reduction or elimination of shade trees, the planting of new coffee varieties, and the application of chemical inputs.

At this juncture, I was a young scientist, freshly minted with a PhD in applied ecology and beginning my tenure as an assistant professor at the University of Michigan. It was the summer of 1989. While teaching a course on tropical managed ecosystems in Costa Rica, I sought field exercises to engage students in the research process. Having focused my dissertation on the role of ants in agroecosystems, I was drawn to the intersection of agriculture and biodiversity conservation. The impact of coffee technification on ant biodiversity was an ideal project for my course.

Together with my students and my colleague and partner, John Vandermeer, I studied the diversity of ground-foraging ants in coffee farms with varying degrees of shade. Unsurprisingly, our findings revealed a dramatic decline in ant species with the elimination of shade trees. But, we wondered, if ground ants dwindled with the loss of shade, what about arboreal ants? To answer this, we surveyed the canopy of two shade trees on a coffee farm. Astonishingly, we discovered 30 ant species and 126 beetle species in one canopy and 27 ant species and 110 beetle species in the other, with only an 18 percent overlap of ant species and a 14 percent overlap of beetle species between the two trees, separated by less than 100 meters. Clearly the technification program was leading to the loss of hundreds of species.

Over the next thirty years, I spent countless hours documenting the intricate ecological interactions on shaded coffee farms. My research journey eventually brought me to the highlands of Chiapas, Mexico, where a dedicated farmer named Walter Peters had established the *Finca Irlanda*. One day, as Walter and I strolled through the lush farm, we stumbled upon a disconcerting sight: a surge of caterpillars voraciously consuming the shade trees. With a calm demeanor, Walter reassured me, saying, "Don't worry, Ivette. The migratory birds are beginning to arrive, and they'll handle this." True to his word, just a few days later, the skies were teeming with migratory birds eagerly feasting on the plump caterpillars. By the next day, the caterpillars had vanished entirely!

This fascinating observation led me to collaborate with Russ Greenberg, director of the Smithsonian Migratory Bird Center, to initiate a research project examining the role of birds in pest control on coffee farms. Over several years, our study revealed a harmonious interplay between birds and bats in regulating pest populations. Birds diligently control insects during daylight hours, while bats take the night shift. Their efforts are beautifully complementary. Our findings also underscored the profound impact of shaded farms, revealing that these habitats foster greater abundance and diversity of both birds and bats compared with unshaded farms. This research highlights the crucial role of shade trees in promoting robust ecological communities, ultimately enhancing the natural pest control mechanisms that sustain coffee production.

During our study on the role of birds and bats on coffee farms, we also stumbled upon what seemed to be a keystone ant species with intricate interactions that effectively control several pests of coffee. We observed that this *Azteca sericeasur* ant engages in a mutualistic relationship with scale insects that extract sap from coffee plants. The ants shield these scale insects from predators and parasites, receiving in return a sugary liquid secreted by the insects. If that was all there was to the story, we might assume the scale populations would skyrocket, decimating the coffee plants.

But it turns out that the ants also provide the perfect nursery for the lady beetles that eat the scales! The ants can't attack the beetle larvae, because they are covered in waxy filaments. Other insects know how to penetrate the waxy covering, but any would-be predators of these beetle larvae are deterred by the ants' aggressive behavior. By balancing out the pressures along the food chain, the ants effectively protect the beetle that regulates the pest population across the entire farm.

As is often the case with science, I was excited by our findings but troubled by a nagging question. I understood that the beetle larvae were protected by their waxy buffer, but what about the adult female beetles that had to wade through a gauntlet of ants to lay the eggs that would become these larvae? This is where things got really interesting. It turns out that a specific kind of fly parasitizes the ants. These flies can locate the general vicinity of the ants from their odor but can attack only a moving ant; stationary ants do not provide them with a target. So when one of these flies arrives in the area, the ants all assume a catatonic posture and remain motionless. This temporary reprieve allows the adult female beetle to approach the plants and deposit her eggs safely.

These elaborate webs of interactions have long fed my curiosity and my soul and stimulated me to reflect more about human–nature interactions and the complex sociopolitical structures that shape those interactions. After all, it was in these large coffee plantations of the Soconusco region of Chiapas that migrant Indigenous workers took the first coffee seeds to bring back to their villages in the highlands and establish new kinds of coffee farms.

As I came to appreciate over decades of research along this fertile coastal plain, Indigenous peasant farmers had assimilated coffee into their traditional systems, crafting new, diverse agroecosystems that married Indigenous practices with a globally connected commercial crop. Forming cooperatives, they sought fair trade certifications aimed at ensuring better trading conditions and sustainability. In Chiapas, this quest for fair trade has intertwined with broader struggles for autonomy and self-determination: The Zapatistas, a revolutionary Indigenous group, have adeptly harnessed the fair trade coffee market, blending economic empowerment with political resistance and cultural preservation.

Ironically, coffee, once an emblem of colonial exploitation, now empowers Zapatista communities to engage in fair trade markets, countering neoliberal globalization and corporate control. By fostering an alternative trade system, they resist commodification and exploitation. Engaging in fair trade also enables them to maintain agroecological practices aligned with their respect for nature and resistance to environmentally destructive practices promoted by large agribusinesses.

The odyssey of coffee, from its humble beginnings in the highlands of Ethiopia to the bustling coffeehouses of Europe, through the horrors of slavery and exploitation in the Americas, and finally to its role in the empowerment and resistance of Indigenous communities in Chiapas, paints a rich tapestry. Along this journey, my minor contributions have focused on that corner of the web of agroecological understanding that acknowledges the complexity of plant and animal interactions and how they intersect with farmers' deep and growing knowledge within coffee's diverse growing environments, continually deepening my appreciation for this remarkable plant and its profound influence across cultures and histories.

Fruit Finding

Eliza Greenman with Bill Davison

One February morning in 2016, I set out on North Carolina's coastal plain. I had received coordinates from Dr. David Shields, a Southern foodways explorer and historian, which led me down a sandy lane to the remnants of a nineteenth-century hog and chicken orchard. This orchard contained the 'Hicks Everbearing' mulberry tree, one I had been searching for ever since reading about how a single tree could produce enough fruit to feed four hogs over the summer. I'm an orchardist who plants trees to help feed livestock, so the chance to see this mulberry was a long-anticipated moment.

I found the stand of ancient mulberries, whose massive, gnarled trunks showed the wear and tear of time. Several of them had lost their crowns to hurricanes that swept through the region. But what made these trees seem so dwarfed was not their lack of crown. It was the 50-foot persimmon tree towering above them.

The old persimmon was perfectly spaced between two mulberry trees, which told me it had been deliberately planted. Yet unlike the famous 'Hicks' mulberry, this particular persimmon was nameless to me.

Growing up in Virginia's coastal plain, I had been taught to think of the persimmon as a weedy tree that attracted varmints, not as a prized resource for feeding livestock. I now realize that my cultural conditioning caused me to underestimate this remarkable tree. After all, its Latin name, *Diospyros*, means "fruit of the gods." My chance encounter sent me on a fascinating journey to understand the saga of this tree: from its long history as an Indigenous staple food, to its key role as an important tree of the enslaved, to its potential future as an agroforestry crop for humans and livestock alike.

The American persimmon, *Diospyros virginiana*, is native to the Southeastern United States, the lower Midwest, and the Mississippi River Valley. Slow-growing, it has two forms: shorter, orchard-like trees with larger fruit, thriving in open fields; and tall forest trees, sometimes reaching 80 feet, which bear smaller fruit. Regardless of size, both types were once incredibly valuable for sustenance.

Indigenous peoples have intentionally stewarded persimmon as an essential carbohydrate. In the fall, persimmons would rain from the trees, their sweet, jelly-like flesh tasting of honey, apricot, and citrus. Too plentiful to eat fresh, they were often dried into bricks or fruit leather for long-term storage. In addition to their high sugar content, persimmons are the most nutrient-dense native fruit in North America, packed with essential vitamins, minerals, and antioxidants.

Persimmons were first reported by colonists in 1541 during Hernando de Soto's expedition to the Mississippi River. When De Soto arrived at the river, he encountered thousands of Native Americans who sent over canoes filled with dried fish and loaves of dried persimmon pulp—a gesture meant to gauge the Spaniards' intentions. Sadly, after receiving the gift, De Soto initiated war with a crossbow attack.

Two centuries later, botanist William Bartram traveled through Georgia and again encountered Indigenous orchards of persimmons, mulberries, and other native fruits and nuts. Growing in fields rather than forests, these trees had clearly been cultivated for better fruit production, as open-grown trees are typically more prolific. When Bartram asked about these plantings, in nearby Creek settlements, people spoke of their value as a crucial food source.

~

Meeting the persimmon towering among the mulberries must have primed my awareness. Later that same year, while I was on a morning walk through the site of an old Quaker village in northern Virginia, my eye was caught again by persimmon. This time it was the orange gleam of their fruit that drew me in. The tree was about 200 feet away, and when I walked up to the fence to get a better look, the trunk prominently displayed a graft mark, telling me that this persimmon once had a name, and someone planted it with a purpose in mind. On old persimmons, graft marks look like someone tied a wire around the tree. Given that this graft mark was about a foot off the ground, I knew the tree had never been used to feed hogs. Their large, clumsy bodies would have quickly snapped the tree off at a low graft union. Was it for humans, then?

This was exciting. I'd never seen a persimmon tree like this, which looked to be grafted more than 170 years ago. The fruits were big, too—about 2 inches in diameter, larger than any other American persimmon I'd seen. Next to this tree was another grafted persimmon, taller but with much smaller fruit, which I thought was odd. If the tree with the large fruit was grafted for humans, why would they bother with also grafting a tree that produces much smaller fruit?

I went home and read everything I could find about this farm and its former owner, Yardley Taylor, a prominent nineteenth-century Quaker horticulturist and abolitionist. Letters penned by Taylor promoted the improvement of agricultural efficiency, to reduce reliance on enslaved labor. One such improvement strategy was grafting and growing superior fruit and nut cultivars to sell in the burgeoning Washington, DC, market in order to improve profits without increasing harvest costs. I thought of the two grafted persimmon trees: one with small fruit and the other with fruits double the size. The two trees, side by side, were the visualization of improvement: One tree could fill a bushel box with half as many fruits as the other. Perhaps these grafted persimmons were part of an abolitionist strategy, a physical representation of how a larger-fruited specimen could reduce the labor needs by half.

I walked by these old persimmon trees whenever I could. One evening, as the sun was setting, the bright orange fruits were radiant as dusk fell across the landscape. I wondered how many local residents had noticed them in passing. To my knowledge, northern Virginia's colonized food culture had never included the American persimmon. Could these trees have remained hidden in plain sight for over a century?

Then it struck me. This farm was a locally known stop on the Underground Railroad. What if these persimmons were part of another abolitionist strategy? One that played to their unfamiliarity among white people. Habitually overlooked in favor of other fruit trees such as apples and pears, persimmons hardly registered in the consciousness of plantation owners. But African Americans knew them well.

In Africa, the jackalberry (*Diospyros mespiliformis*) is a widely known persimmon species and one of the most commonly grown trees in the continent's west and central savannas. The jackalberry's fruit, one of the oldest known human foods, is cherished for its diverse culinary uses, and the bark and leaves are used to produce various medicinals. Enslaved Africans, encountering the American persimmon within its native range, recognized its similarities to the tropical jackalberry and discovered that it could be used in much the same way. They ate foraged persimmons fresh, used them as a trap tree to hunt wild game, and preserved them through drying, baking, and fermentation. Dried persimmons could be mixed into cornmeal to make a hearty bread or brewed into a refreshing, low-alcohol beer. These recipes were created based on the ingredients on hand.

Black cooks in plantation kitchens were adept at transforming traditional European recipes, such as British puddings, by infusing them with bold flavors and local ingredients. Wild fruits and nuts were frequently used to elevate simple, often bland dishes into rich and vibrant creations. Spices such as cinnamon, nutmeg, and clove— central to African and Caribbean culinary traditions—added warmth and depth, turning these adaptations into entirely new recipes.

It is possible that persimmon pudding, a dessert gracing church potlucks and Thanksgiving tables throughout the Southern Piedmont and across the Midwest, was the result of a Black cook's inventive riff on English-style puddings, combining the native fruit's natural sweetness with the distinctive spices of African American cooking. Somehow, perhaps because of the fall abundance of this fruit paired with a flexibility in ingredients, this dessert has managed to keep the once mighty persimmon in American foodways. In many places, the dessert is becoming a victim of attrition, as younger generations reject it from their holiday menus. But this is not the case in southern Indiana, where persimmon pudding is still king.

～

In the fall of 2022, I visited the Mitchell Persimmon Festival, home to the only celebration of American persimmon in the United States. The heart of the festival is the persimmon pudding contest, which brings the community together in celebration of this American twist on English pudding. I watched as three generations of the same family filed in to enter their pudding into the annual contest, with hopes of becoming the next Puddin' Champ. With hundreds of puddings lined up on tables, tension filled the air as the judges scrutinized the 4×4-inch pudding squares.

Back in the 1950s, the Mitchell Persimmon Festival drew crowds upwards of 40,000 people. Conference organizers seized the opportunity to use the festival as a means to locate and propagate the best persimmon trees. A $15 prize, with the enticing possibility of an additional $10 for an exceptional variety, motivated growers from near and far to bring their finest persimmons. Soon a collection of the top varieties spread to a diverse network of researchers, nursery owners, and enthusiasts across the country.

In 1959, these festival favorites (and many other persimmons) were grafted at the St. Elmo, Illinois, experimental orchard of Jim Claypool, who recognized the potential value of the fruit pulp for pudding and other culinary markets. He knew that all the varieties available to him were selections from the wild, and to breed better pulp-producing persimmons, he had to start making crosses. Over his lifetime, Claypool made 1,500 crosses, the most promising of which were crossed further by his protégé, Don Compton. These crosses produced four promising new varieties for persimmon pulp.

But it took nearly sixty years for these traditional breeding and selection methods to yield results. This long timeline and the lack of institutional support hindered progress for American persimmons. Today, advances in genetic tools—such as affordable genome sequencing and molecular markers—can help accelerate breeding timelines from decades to years. With these innovations, there has never been a better time to focus on orphaned tree crops.

∼

Inspired by all that I had learned about persimmons and determined to get to know this prolific tree a little better, I made a pilgrimage to

Don Compton's orchard to collect some seed. It was unusually hot and humid for mid-September in southern Indiana, with temperatures in the mid-90s. I had arrived the afternoon before harvest to clear the orchard floor of previously fallen fruit, ensuring that only freshly dropped persimmons from select trees, such as 'Dollywood,' 'Killen,' and 'Wannabe,' would be collected. This way, I could accurately track which fruits had fallen overnight from which tree.

The next morning, just before sunrise, I began gathering the freshly fallen fruits, tossing them into labeled buckets with the mother tree's name. Crawling from tree to tree, knees covered in orange goo, it took me three hours to harvest six rows of persimmons. Once finished, I took all the buckets to the barn, where I mashed the persimmons into a pulp using a garden rake. I then covered the buckets (to lessen yellowjacket pressure) and allowed the contents to ferment so I could more easily extract the seeds from the pulp.

I repeated this harvest–mash–ferment pattern for the next three days and eventually had 1,700 pounds of pulp, a heavenly smelling sticky goo, ready to process. I dumped the bucket of pulp onto a screen and pressed it through the mesh, leaving the shiny black seeds on top. After days of harvesting and processing, persimmons had permeated my clothes and skin. I was worn out and deeply content. I now had 75,000 persimmon seeds that represented the hopes and dreams of generations of tree breeders.

Among these seeds are the genetics of persimmons enjoyed in the Mitchell Persimmon Festival's contests, selected through the work of Claypool and Compton, and before that cared for by generations of people beyond my own cultural and scientific tradition. When these seeds become seedlings, 5,000 of the best performers will be transplanted into the Savanna Institute's new breeding orchard in Spring Green, Wisconsin. Here, plant breeders, along with Bill Davison of the Savanna Institute's Commercialization Department, will select for traits that make persimmons viable in modern agroforestry systems and food and beverage markets.

It will take a team effort to restore American persimmon to its rightful place in our everyday diets. We'll need growers, pulpers, and chefs to pitch in to help us improve the fruit's fresh eating and processing quality, and we'll need livestock farmers to experiment with integrating these trees into pastures as a supplemental source of animal nutrition. As with any relationship that has lapsed, we imagine there will be some bumps along the way. But as we reacquaint ourselves, this resilient tree can once again connect people and livestock with the land and provide a valuable native staple crop.

Elderberry Season

Jesse Smith

In my California childhood, it seemed everything was in motion. From the geology to the hydrology, the climate to the demographics to the politics, the Santa Barbara of my youth was a living canvas of change and growth that felt intensely alive. The salty seas of the Pacific Ocean. The warm breezes that drifted down from the mountains. Raised by a strong mother within a supportive community, I inherited a fusion of Afro-Caribbean and Euro-American lineages, which deepened my reflection on how concepts of place and identity evolve across generations.

Much as I loved my coastal birthplace, I realized early on that true appreciation for life's diversity requires exploring beyond the familiarities of home. This realization propelled me to metropolitan hubs like Los Angeles, San Francisco, and London, where I pursued print production, graphic design, photography, and product design. I found myself fascinated with the impact of storytelling on how humans interact and experience the world around them, and I spent several years working as a communication consultant. Later, I came to reflect on how these early years learning communication design would stand me in good stead as I steered my career toward food and agriculture.

As I merged my professional expertise in design with a longtime love of food, my passion and experience in regenerative agriculture grew, and I embarked on a decade-long journey into food system design and land stewardship, guided by the principles of permaculture, holistic management, agroecology, and organic agriculture. Growing food inspired me to develop my passion for the culinary arts and to find creative ways of bringing people together around our common bonds to land. Along the way, I learned how to build enterprises and educational programs, and I was astounded by how many people were eager to get their hands in the dirt. I realized that my pull toward land stewardship reflected a larger yearning, a reawakening to the intrinsic connection between humans and the natural world.

In my current role, as director of land stewardship at White Buffalo Land Trust, I support efforts to advance regenerative agriculture that restores soil health, water cycles, biodiversity, and human health. We're putting these principles into practice on our 1,000-acre ranch by developing managed grazing and agroforestry systems while monitoring the impacts of our management on the land and sharing what we've learned through educational programs and trainings. In tandem, we're also developing products like a persimmon vinegar and a dried beef snack to grow the marketplace for regeneratively raised foods. Our vision is expansive: to diversify agriculture by integrating a mix of perennials, annuals, and animals into abundant and resilient systems that maintain ecological balance and enhance viable land-based livelihoods.

We have identified numerous crops and systems that provide opportunities to support these efforts. However, the standout in the palette of plants and animals is the western blue elderberry, *Sambucus cerulea* (*Sambucus nigra* ssp. *cerulea*). This native species thrives under the variable climatic conditions of California, demonstrating remarkable versatility in providing food, fiber, medicine, and material. At the same time, western blue elderberry supports multiple forms of biodiversity: Previous research has demonstrated that adding these plants to farms as hedgerows increases populations of pollinators and insects that eat crop pests, while providing habitat for native songbirds. Meanwhile, the roots of this hardy plant stabilize fields and stream banks, enhancing soil health and water cycle function.

It's incredible just how many places you can bump into an elderberry. This plant grows among grasslands and along riparian corridors, hugging oak woodland edges and dotted among coastal sagebrush. You can find it all across the temperate zones of North America and to the far reaches of Europe and beyond. This ability to thrive in such a wide range of places makes elderberries not only an exemplary candidate for regenerative agriculture but also a symbolic reminder of nature's ingenious resilience.

This plant has been an integral part of my land stewardship journey from the very beginning, though I didn't realize it at first. In 2013, my wife Ana and I, along with family and friends, launched a collective of local food businesses on a patch of farmland just down the coast from Santa Barbara. Together, we crafted organic artisan cheeses, raised heritage breed pigs, turkeys, and chickens, and managed avocado, apple, and persimmon orchards. We also developed an intensive annual market garden alongside microgreen and culinary mushroom operations. This venture served as a profound learning experience for our young family, illustrating both the opportunities and challenges of diverse and integrated agriculture. Throughout these formative years, which also welcomed the birth of our eldest daughter, Phoenyx, in 2013, our lives were deeply intertwined with the rhythms of a vibrant, living landscape.

Each day, as I traversed the path from our front porch, I brushed past the branches of a mature elderberry tree whose limbs stretched out invitingly. From my deck, I could easily harvest clusters of the powdery ripe blue berries, constantly reminded of the tree's capacity to offer medicine, shelter, and beauty year-round. Yet this relationship also taught me the importance of attentive care. The elderberry was not meant to thrive unaided; it needed regular pruning, tending, and harvesting to reach its full potential. This enduring lesson has stayed with me: In nurturing nature, we enable it to nurture us, highlighting our vital role in the reciprocity of life.

Many times, disturbance precipitates significant change. At the close of December 2017 and the outset of 2018, our community was shaken by back-to-back natural disasters. First came the Thomas Fire, which burned more than 280,000 acres for several days, threatening life and limb across the coastal mountains behind Santa Barbara. Our family evacuated on the first night of the fire and once again nearly a week later as it moved north and threatened new areas. Having lived through multiple major incidents, including the loss of my family home in the Jesusita Fire of 2009, we had become somewhat accustomed to these evacuations. However, the impact of the Thomas Fire was exacerbated by the subsequent mudslides and debris flows triggered by intense rainfall in early 2018. These events destroyed or damaged more than 500 structures and claimed the lives of twenty-three people. These consecutive catastrophic events profoundly altered our community's perception of our environmental and social vulnerability. When White Buffalo Land Trust was founded in March 2018, there was a deep resonance between our mission and the community's urgent need to invest in the foundational health of our land. To prepare for the unpredictable future of climate change, we recognized the necessity to take major steps in transforming how we engage with our living and dynamic landscape.

The integration of elderberry into our farming practices at White Buffalo Land Trust transcends conventional agricultural productivity. It's about revitalizing and nurturing the profound cultural and ecological connections that perennial climate-adapted plants inherently support. Through initiatives that engage the community—such as workshops, educational tours, and collaborative projects—we emphasize the elderberry's vital role in our shared environmental heritage and its capacity to lead regenerative agricultural practices. These efforts go beyond simple cultivation; they aim to create a living library of plant-based knowledge that enriches the ecological and cultural fabric of our community.

When people see these farming systems in action, my hope is that they will recognize and reward the benefits of regenerative agriculture in its many forms around the world. We all eat; it's the great connector. However, as I have seen interest in regenerative agriculture grow, so too have I seen the pitfalls of bringing old ways of thinking and doing business into this emerging field. The habit of extracting value before value has been created is a hard one to break. As a result, much of my work these days is focused on finding ways to work in reciprocity with other stakeholders, avoid transactional relationships, and make sure we lead with our principles and infuse them throughout the work. Our aim is to catalyze a wider transformation in our society that uses agriculture as a forum for engagement, a place for connection, and an arena to find a shared language.

In 2023, our organization was honored to receive a grant from the US Department of Agriculture to support our comprehensive efforts in growing, processing, researching, and marketing the western blue elderberry. This initiative, known as the Elderberry Project, represents a collaboration between multiple regional partners orchestrated by White Buffalo Land Trust. Partners include the Community Environmental Council, University of California Agriculture and Natural Resources, the Santa Ynez Band of Chumash Environmental Office, and numerous land stewards, processing partners, and businesses eager to incorporate this crop into their products.

As excited as I am about the potential of elderberries on a larger scale, few things make me appreciate this plant more than navigating cold and flu season as a working parent. Over the years, it has become our go-to remedy, providing gentle and comforting reassurance that natural remedies and health tonics can be made locally and at low cost. The way elderberry seamlessly fits into the lives of my children and family, offering its medicinal benefits with such ease, mirrors its ecological adaptability. Whether used in soothing syrups during cold seasons or as an elderflower liqueur my wife makes each spring for a refreshing addition to lemonade and a summer spritz, we seem to be able to enjoy the benefits in sickness and in health. It lends itself to various culinary and medicinal applications with grace, infusing confections, supplements, and even celebratory beverages. I look forward to the next time I'll be able to smell the sweet and rich aroma of the herbs and spices reducing in a pot of elderberry juice. The silky texture dripping off the back of a spoon and the stain-soaked fingertips after a day of making our winter's cache of elderberry health tonic. As much as I see the potential of elderberry-based products made to share with the world, there is a special place in my heart for the unique recipe and care that can come only from one's own kitchen.

As I look to the future, the lessons I draw from the elderberry are more relevant than ever. In a world facing environmental uncertainties, cultivating plants like the elderberry, which thrive across varied settings and offer multiple benefits, becomes crucial. It is not just about fostering biodiversity but about adapting our practices to reconnect our human lives with the living cycles of Earth, both responsibly and creatively. The elderberry, with its humble yet profound versatility, continues to inspire and challenge me to think differently about my role in nature.

Hazel-Bush and Willow

Fred Iutzi

I don't know how to define agrarianism, but when my firstborn child came home from the hospital in 2012, the first thing I did was to walk him to the front windows and look back out on the world we had just come from. These glass panes first felt the eyes of our ancestors in 1894, when my great-great grandparents (that's three greats to you, son) William and Alice Lambert built a house on Section 1 of Rock Creek Township, Hancock County, Illinois—west central Illinois, corn country—and installed three whopping big sash windows. The reason it's so cold standing here on this January day, little one, is that my great-grandparents (two to you, son) reinstalled those silly windows, every bit of 10 feet tall, into 8 foot ceilings when they built the new house, and every molecule of warm air is disappearing up into those ceiling pockets. But cuddle close and look outside—you'll run and play in that yard, like generations before you. South of the lane is where your grandpa will plant corn in the spring. I'm not all that old, but I can remember last time there were oats in that field, and, wow, we had one heck of a straw fire when your great-grandpa got too ambitious with the hay rake. You and I might plant oats again there someday. Look past that field, and past that, and past that—out into the wide open of the old Hancock Prairie.

I don't know how to define agrarianism, but I'm pretty sure we were being agrarian pretty hard in 2000 when a bunch of undergrads were bumping along in the back of Wes Jackson's pickup truck across the intact (sort of) prairie of the Flint Hills of Kansas, imbibing the fundamental difference between that plant community—made up almost exclusively of perennial plants, spanning a distinctive range of functional ecological niches and life history strategies—and the annual monocultures that cover our grain crop landscapes. The key to the agricultural sustainability puzzle, to the reset and renewal of our agricultural traditions, must be the prairie—that proxy system for our current fields and pastures from back when the world was a little newer.

I don't know how to define agrarianism, but in the fall of 2014, when my son was two years old, going on three, we had a good and long apple season at the Lambert farm. For weeks, when we got home together from work and daycare we would bypass the house, bypass the play area outside the door, and walk back to the far corner of the yard. I'd grab an apple for each of us, from one of the three scraggly apple trees my uncle planted there a couple decades prior, and we'd amble. Often we'd leave the mowed area of the yard and cross into far country for a two-year-old on foot, into the Osage orange–lined pasture along the creek or down the rows of tall corn in the field—I have a photo of him there, apple in hand, adventuresome smile on his face.

I don't know how to define agrarianism, but maybe my great-grand-parents—two sets of 'em, the Lamberts and the Stevensons—did back in the 1910s and 1920s when their farms were a go-to destination for University of Illinois ag professors looking to bring their students to see the latest in green manures and crop–livestock integration and soil erosion–fighting terraces and rural organizing. And maybe my parents could define it in the 1980s when they told me I could do and be whatever I wanted to, and in the event that had something to do with farming, here's your history and here's the history of every single part of this land. And I'm sure I was filled with the spirit of it the first time I was entrusted with the awesome responsibility of running the corn planter in the 1990s, and I'll bet I would have been overconfident enough to spout off about it in a sustainable ag graduate seminar in the early 2000s.

I don't know how to define agrarianism, but I have been propelled by it as long as I can remember. There are lots of problems to solve in dominant culture American agriculture, lots of wrongs to right. Despite the best efforts of generations of good farmers and good scientists, the continent is still washing into the sea, a large measure of nutrients and carbon best left in the soil are instead in the rivers and the sky, and a litany of race, gender, and class injustices are still ongoing. My own work in sustainable food systems and rural cooperatives and perennial agriculture is motivated by an abstract desire to do good and a recognition that it will take clear thinking and a rigorous effort to find and apply the right technical and social means to solve problems. But really? What draws me in and keeps me there and moves me forward is that emotional and aesthetic and historical connection around family and crops and livestock and farms. An affection so deeply felt that it was instinctively the first way I ever attempted to orient my children to the world.

Back on the evening of August 11, 2021, I was putting together slides for an interview presentation for a job at the Savanna Institute. It was my first foray into agroforestry, but I had the agro part down cold and had dealt with perennial crops a little bit in my time. They wanted to know something of my background and motivation, so obviously my slides about the family farm went in, complete with scanned images from the yellowed pages of old local histories. I had read these stories many dozens of times and used this slide in probably ten talks already, and I had the key sentences outlined in red, to make sure nobody missed them. But I had never applied for a job at an agroforestry organization before—which was probably the reason I read a little further that night, to a tidbit about the establishment of this farm that I'd somehow never seen: "The country was covered with hazel-bush and willow, and it required all the power of five yoke to tear up their roots." I didn't skip a beat in extending my red outlining on the slide image, to make sure that my agroforestry job interviewers would know how deeply engrained woody plants were in my consciousness. But I was completely astonished.

A key idea from American agrarian writing of the last century is the "conversation with nature." Wes Jackson and Wendell Berry suggest that dominant culture farmers ask, of any given farm, "What was here? What will nature require of us here? And what will nature help us do here?" These questions are a good source of inspiration and method for any paradigm of sustainable agriculture community but are particularly special to the perennial agriculture movement. Because for most North American farms, what was here before the plow was an ecosystem built around a diverse perennial plant community, with a rainbow of regional and local specifics. It would be a misstep to presume that the precise history of the land must be rigidly deterministic of what comes next, but it's invaluable information.

And yet my agrarian gaze had skipped over "hazel-bush and willow" dozens of times. I thought I had been an effective listener in my conversation with nature. Clearly I had not. The two species from the text aren't little wisps: American hazelnut has heavyweight economic significance in agroforestry conversations, and willow is literally a tree (or at least it can be if not burned or clipped constantly). I had been missing out on important mental agility in understanding the ecological situations around me and what could have been a deep connection to a huge swath of wild woody plants and agricultural tree and shrub crops—because *of course* woody plants are part of an Illinois prairie landscape.

To figure out why my agrarian conversation with nature went astray right away in the listening phase, at least where woody plants are concerned, I suspect I need to have a conversation with culture. And in a story about a farm in Illinois, written in English by a native English speaker who is American, White, and of entirely European ancestry, a careful reader may have detected clues about which cultural tradition we're about to chat with.

When I held my newborn son up to those ancestral windowpanes in 2012, the lens we gazed at the world through was an old one: The tiny narrative journey we took from house to farm field traversed a cultural path already well worn. The farm field is the eponym for the whole Western cultural proposition of agriculture, and what we physically do and do not find there can powerfully shape our perceptions of what is appropriate, what is laudable, and what is possible. That there is little recent tradition of trees in the part of the Midwestern US crop and livestock farm that we call "fields" looms large for how we discuss the future.

Meanwhile, look for the trees that *were* present in my narrative. Picking and eating apples with my son in the fall of 2014 is a cherished memory for me but not one I experienced as agricultural. The Osage orange trees I mentioned were also beyond my sphere of agriculture. These hedge trees that line the creekbanks are in the least managed, wildest part of the land, seemingly outside the farm field.

In short, the Western cultural frame imposes boundaries on this agrarian scene. The agrarian gaze perceives three zones, in order from safest and most controlled to wildest and least controlled: yard, garden, and orchard; crop field and livestock pasture; and forest, woodlot, or whatever passes for wild and unmanaged land. Two of these zones, the least and most wild, are expected to contain trees but are emphatically not part of the golden-hued heroics of the agricultural endeavor. The middle zone is where the farming happens, but it is often implicitly defined as not containing trees—setting the stage for my surprise at woody plants popping up in my family farm's ecological origin mythos.

To solve problems of sustainability, we are well served to include rather than exclude a full range of solutions, and to solve problems of justice we are well served to include rather than exclude a full range of people and perspectives. Whatever worldview one holds, agrarian or otherwise, attention to good boundary work is needed.

First (and maybe most) we need to *surface* boundaries, so they don't catch us by surprise or trip us up. What are we hearing, and what are we walling ourselves off from hearing, when we listen to nature and to each other? Second, in many cases we need to *soften* boundaries so they don't stifle us and prevent us from acting on what we learn; some of us need a refresher in how woody plants played a key ecological role in the prairie from which our agriculture was carved. Third, we need to *honor* boundaries and the categories they define in ways that are constructive. There is a great deal of tree crop knowledge and savvy to recover or to relink that is coded according to categories that have been undervalued or othered. Conversely, honoring boundaries sometimes means taking care not to automatically appropriate that which one has recognized as being of value. And finally, we need to *cultivate* the boundaries. What flourishing and productivity can we bring from the edges between ecosystems, both biological and cultural?

I have come only recently to agroforestry, and it is challenging, but it is a particularly satisfying challenge to take on in the context of all the forms of boundary work to be done and the expanded range of tools that tree and shrub crops bring to agriculture. Agroforestry farmers, extensionists, and researchers are co-creating and co-implementing alley cropping and silvopasture systems, as well as woody edge-of-field practices, that diversify and intensify other forms of annual and perennial crop production. One particularly promising woody perennial oil and protein staple crop, well adapted to Midwestern landscapes and to the future US food system, happens to be the American hazelnut.

I still don't know how to define agrarianism, but in 2015 my son was finally big enough to cross the boundary and clamber down the tree-lined creekbank on the Lambert farm, and that summer he soaked up ideas like a sponge. I can't remember the specific piece of beyond-his-years technical fact he piped up with one particular day that prompted me to say, "Wow, you know a lot about farming." But I remember his response very precisely. After a thoughtful pause he said, "No. I know something about farming."

I don't know how to define agrarianism, but I know something now about how it has defined me. As long as this particular form of meaning-making and joy-finding energizes my actions, it's vitally important that it also energize my looking and my listening. I feel a little awkward about taking so long to notice the hazel outside my fore-bears' back door, but I'm glad I know something more now about the past, and perhaps future, plants of the prairie and the field. And I'm keeping an ear out for whatever comes next.

∿

This essay is dedicated to Catherine A. (Barr) Turney, 1842–1881, who poured out untold love and toil in a place that is very special to me, only to die young and far too anonymously.

II. GRASSLAND

COMPRISING SOME 10,000 SPECIES, the grass family (Poaceae) is the fourth largest plant family in the world. Before mechanized agriculture, grassful landscapes covered nearly 40 percent of the earth's surface. Hotspots of biodiversity and clean water, grasslands are widely considered the birthplace of humanity. And human communities have flourished in grasslands, from the African savannas, Australian downs, and Eurasian steppes to the prairies and pampas of the Americas. Yet nearly half of the world's grasslands have been lost in the past few centuries, many of them converted to cropland. Meanwhile, in the absence of Indigenous land management practices, much of the grassland that remains has been overcome by nonnative species and encroached upon by woody plants.

The North American Great Plains rank among the most threatened grasslands on Earth. Landscapes of shrubs and grasses once covered a billion acres across what is now the United States, from the humid Southeast to the arid Southwest. But more than 34 percent of these lands have been converted to alternative uses, including 96 percent of the tallgrass prairies that previously stretched across the US Midwest. Violent dispossession of Indigenous peoples wreaked havoc on these landscapes, as did the intensive annual agriculture and sprawling development that followed. And yet, despite this grim history, grasslands nonetheless remain among the most ecologically intact landscapes on the continent.

Whether you see prospective meals in such landscapes depends on how well you know them. Indigenous communities have long harvested a variety of grassland plants and seeds, many of them perennials. In California, where grasslands covered 20 percent of the state before 1850, grassland foods have traditionally been harvested everywhere from the coastal prairies along the Pacific Ocean to the high mountain meadows of the Sierras. In the spring, Indigenous Californians knew it was time to dig edible bulbs like soaproot before they flowered. In the summer, they used sophisticated baskets to collect seeds from plants such as California buttercup, gently beating the cluster of food-bearing flowers in a way that ensured only ripe seeds were collected and that some fell to the ground to sow future harvests.

Across the Great Plains, the tuberous roots of tipsin (prairie turnip, *Pediomelum esculentum*) have provided a reliable food for Indigenous communities. Tipsin roots can be dug and eaten fresh or dried and preserved for later use, and they are quite nutritious. Although gathering and consuming tipsin reduces the number of mature plants in that season, the "prairie turnip paradox" names the increased abundance of tipsin in landscapes over time that can come with caring human harvesters using some traditional methods.

Many Indigenous communities have also tended grasslands with carefully timed fires, which not only stimulate human plant foods to grow but can bring tender shoots up earlier in the spring, when bison or elk need them most. As Robin Wall Kimmerer writes in *Braiding Sweetgrass*, "sustain the ones who sustain you and the earth will last forever."

In recent decades, this lesson has taken hold in the regenerative ranching community, with many graziers proudly dubbing themselves grass farmers. Livestock may be what they sell, but the health of their grasslands is the main thing they manage, in partnership with their animals. These ranchers are learning from bison and their stewards, rotating their cattle (and sometimes goats, sheep, or chickens) in patterns that mimic the movements of native herbivores. When the animals graze, it stimulates grasses to replace the lost foliage. When the animals move on for an appropriate period of time before returning, the grass has time to grow, with a little help from the fertilizer the animals leave behind.

Some crop farmers are restoring grasslands into agricultural landscapes too. Grassland buffers at the edges of fields and pastures have become standard recommendations for keeping soil and nitrogen out of waterways. And as you'll read about in this section, a group of Midwestern farmers are now breaking up large corn and soy plantings with prairie strips: blocks of native vegetation that range from 30 feet to 120 feet wide. We won't spoil that story for you, but rest assured that both the birds and the bees approve.

Perennial grasses are remarkably efficient at scavenging nutrients from every segment of the soil profile, and they hold onto them too. One study in Iowa analyzed the impact of perennial strips on mitigating nutrient runoff and found that crop fields without strips lost eight times more nitrogen and 35 times more sediment. Another study assessed grassland buffers at the edge of a grazing system in Missouri, which reduced nitrogen loss by 68 percent relative to grazed land without a buffer. Grasses aren't just keeping all that scavenged nitrogen out of the watershed; they're using it too! This is one reason they grow so robustly without fertilizer or other nutrient inputs.

Evolved to thrive in dry places, grasses use water very efficiently and have developed the capacity to die back in drought times and resprout when moisture is more abundant. They also have an uncanny ability to adjust and adapt to environmental change, a quality scientists refer to as ecological plasticity. Recently, a team of scientists suggested grasslands might actually be the preferred form of perennial plant cover for sequestering carbon in California. Not because they store more carbon than trees (they don't) but because they store a higher percentage of their carbon underground, where it is protected from wildfire or other disturbances. Just walk around in a recent burn area, and you will see grasses resprouting.

Grasslands around the world have been fragmented. But people are working to stitch them back together: protecting old-growth remnants, reestablishing prairies and pastures, and connecting the pieces so creatures can migrate and knowledge can travel. Across the Great Plains, Indigenous communities are restoring prairie foods, including a wide variety of nutritious plants as well as the iconic American bison. Meanwhile, across Midwest row crop landscapes, farmers and researchers are collaborating to reintegrate the pastured livestock that once dotted family farms. As more patches join in this growing grassland quilt, an increasing number of people are seeking out meat and dairy from animals raised on grass. Now that "grassfed" has entered the lexicon, perhaps we are ready to more fully embrace the ways in which these landscapes nourish us.

Landscapes of Abundance

Rosalyn LaPier

A few years ago, I spent a year as a visiting professor at the Harvard Divinity School, where I taught about and did research on Indigenous religion. They had an engaging and vibrant academic community, and most days people met in the school cafe for lunch or coffee. I was surprised by the number of times either faculty or students asked me whether Indigenous religions had sacred texts, the way other religions do. I was always taken aback and replied, "Of course they do. They are just not usually written down in the way you would think, in a book or scroll."

I started tinkering with the idea of how to articulate in writing how differently literate peoples kept "sacred texts."

It was not until I got a call from Tiya Miles—a new professor of history at Harvard, who wanted me to contribute a short essay to a series she was organizing—that I decided to finally write down what I had been thinking about. I originally titled the piece "Land as Sacred Text," but it became "The Land as Text." It was published in early 2023 in the *Journal of Environmental History*, along with essays by Tiya Miles, Lauret Savoy, and Connie Chiang.

In "The Land as Text" I tried to explain how the Blackfeet embedded their history and sacred knowledge within the landscape, within places and place names connected to stories, songs, and rituals. And I shared how my grandparents, particularly my grandmother, would recite dozens of stories each time we drove from Browning to Great Falls, Montana, to go shopping.

Each time we traveled they looked out onto the landscape, seeing a hill or indentation, a river valley or grouping of trees. For them the land was a text. They could see its pages and start reading it like a book. To my grandparents, there were no blank pages or empty spaces, just a book filled with stories.

I learned from them that the Blackfeet landscape was full of humans and nonhumans, monsters ready to kill you, heroes waiting to save you, and Deities who may or may not be interested in your continued existence. It was so full of stuff that you could not even hope for solitude. Even if you were captured by the old grandmother and thrown into her cave made of boulders (one of my favorite stories) you were not alone, because there was always going to be someone else there, too, hiding in the dark.

This is the world I grew up in. A place crowded and overflowing, bursting at the seams with people and activity and action. This is the world that influences my outlook on life today, my relationship to the land, as well as my advocacy for and writing about the natural world.

I want to begin by pointing this out: The Blackfeet lived in a world of abundance. Not only was the landscape full of stories. The land was also full of food: animals to hunt and plants to harvest, process, and eat. I want to challenge the common stereotype that Indigenous peoples were always on the verge of starvation or living a subsistence lifestyle, just relying on "what nature provided." In fact, Indigenous women were actively managing the land to create landscapes of plenty. The Blackfeet have very few stories that tell of scarcity. One of the few is the story of the Iniskim, or buffalo-stone, where in a time of a bison shortage a supernatural entity sought out a Blackfeet woman to teach her how to entice the bison to go anywhere she wanted them to go. Even in this instance, the moral of the story was about how to manage and maintain abundance. This is why, when I encounter narratives that erase Indigenous peoples off the land—or define Indigenous lands as "wild" and "empty"—I push back.

Rain

During my grandparents' childhood, near the end of winter, the Blackfeet held a ceremony that marked the beginning of the new year. They stored dried berries, dried meat, and other dried plant foods and teas to use at this ceremony. These dried items were what remained of their harvesting and hunting from the previous year. At this ceremony they did not serve any fresh plant foods or meat. But this was intentional and part of its purpose.

When my grandparents were children, the Blackfeet religion was outlawed on the reservation by the US government. To practice their New Year's ceremony, their family and others moved it to Easter Holy Week. At that time both of my maternal grandparents attended Holy Family Mission, the Catholic mission school on the reservation, and Easter Holy Week was one of the few times that the Catholic Church allowed parents to visit their children. My grandparents' family set up camp next to the school on the Two Medicine River and held their ceremonies every night alongside the daily Easter Holy Week activities. This adaptation away from following the cycles of the natural world to following the Catholic liturgical calendar was a small price to pay to continue practicing their religion.

This New Year ceremony celebrated, recognized, and acknowledged the return of rain to the prairies and the importance of rain to replenish the life of the soil, plants, animals, and humans. Without rain, nothing could live on these arid and semiarid regions of the northern Great Plains. As part of their ceremony and celebration of the return of rain, each Blackfeet person took individual dried serviceberries, held them up to the sky, prayed to the Deities, and then "symbolically" planted them in the ground. This ceremony had been conducted year after year for thousands of years.

Serviceberries

The early recorders of Blackfeet life noted that while no one had made an "exhaustive study of their food" and although they had "no traditions of agriculture," the Blackfeet "consumed a considerable amount of vegetable food." And "the serviceberry was the most important vegetable food" eaten by the Blackfeet. Of course, if you had asked my grandmother, she would have told you this.

I would agree. Based on my experience apprenticing with my grandmother and oldest aunt learning about Blackfeet plant knowledge, and my now years of experience, our historic diet was largely plant based, complicating the narrative that our main source of food was bison. The Blackfeet historically ate a wide variety of root vegetables, fruits and berries, seeds and nuts, and leafy greens. Plus plants were used for teas, seasoning, and preserving foods.

When my mother was growing up, she remembers picking gallons of serviceberries each summer. It was a big family activity that gathered everyone together. If you don't know, serviceberries are about the size of a blueberry, and they grow on perennial bushes not much taller than most people or small trees. And to preserve their berries the Blackfeet historically dried them.

My mother's family usually dried their berries on large canvas tarps, typically from an old canvas tipi or a tent, and they laid these on the ground or sometimes on the roof of my grandparents' small house. After the berries dried, they would shrink down (smaller than most raisins) into hard little pebbles, which is why the Blackfeet called them Okonoks or rocks.

They gathered enough serviceberries to fill several 50-pound cotton flour sacks after they were dry. Then they ate them all year long, through the winter, spring, and summer, until they could harvest them again at the end of the summer. They ground them into a flour to add to other foods, or boiled the whole dried berries to make hot soup. By the time I grew up, we relied less and less on eating serviceberries every day, and when we picked them we often used a combination of drying or freezing to preserve them (because folks had electricity then).

Even beyond their importance as a food, serviceberries play a central role in Blackfeet cosmology and history. The inner bark of their branches was used as medicine, and their wood was used for making tools. Very small sharpened serviceberry twigs were used to pierce the ears of babies, and straight long branches were valued for making sacred pipes. When the plants bloomed in the spring, it was time to call people together to plan a ceremony. And throughout the year, the dried berries were made into a soup served at almost every religious ritual.

Singing

The one domesticated plant that the Blackfeet grew was tobacco. In the early spring, not long after their New Year's ceremonies, the Blackfeet cleared plots along creeks to create a garden. First, they went to the serviceberry bushes and cut back the dead wood or branches they thought were overgrown. They also coppiced or pruned the bushes, which would stimulate new growth or create straight branches useful for particular arrow shafts or pipe stems. Next, they took the cut branches to the clearing where they planned to create their garden plot and burned them. After that, the Blackfeet raked their garden plot, mixing the soil with the ash. Before they planted, a religious ritual was held to honor the Deities and ensure a healthy crop. They mixed their tobacco seeds, which are very tiny—much smaller than poppy seeds—with a mixture of ground-up dried serviceberries, animal manure, and water. The Deities told them, "This will enrich the soil."

My grandfather used to delight us children with stories of the Deities who took care of the tobacco gardens. The Deities enjoyed singing, and they stepped on insects in the tobacco gardens while dancing. By my childhood, tobacco was store bought, and no one grew it. But I learned that the Deities who helped take care of plants, whether domesticated or cultivated, were still important to Blackfeet society, even though at that moment they were not being called on to serve a duty.

When my grandmother grew up, she learned to sing the songs that these Deities, who also helped serviceberry bushes thrive, enjoyed hearing. The songs lulled insects to sleep, and the Deities would brush them off the serviceberry bushes. When my mother grew up she often heard her mother and aunts sing these songs as they picked berries.

Once I was doing research at a natural history museum and I came across the fossils that the Blackfeet used with their singing to these Deities. The museum called them "worm stones." But they are fossils of ancient ammonite, not stones. And similar to the "buffalo stones," "worm stones" were used to change the behavior of insects in the same way "buffalo stones" were used to change the behavior of bison.

Fresh

I am often asked which came first: ecological knowledge and then religious practice or religious practice and then ecological knowledge.

Hmmm . . . I am not sure if it matters. What I do know is that the Blackfeet understood ecological processes. They understood that insects were pests that could do harm to domesticated and cultivated plants, and this knowledge was incorporated into their cosmology and religious practice. They understood seasonal cycles, climate and weather patterns, plant and animal behavior, and these were incorporated into their religious practice. But they also knew how to change the landscape they lived in. They transplanted, pruned, and coppiced, collected and planted seeds, used fertilizer, and set carefully timed fires, increasing the abundance of plant-based foods and medicines.

Toward the end of summer, when the serviceberries are ripe, the Blackfeet hold another ceremony. It marks the end of the summertime and rain and the beginning of preparation for a long winter. And unlike the ceremony held at the end of winter where all the food is dried from the previous year, at this ceremony all the food is fresh—with fresh berries, meat and wild mint tea, all recently harvested. The Holy Woman who leads the ceremony fasts, not eating or drinking for several days. She is only allowed to drink what my grandmother called "muddy water," or water infused with earth minerals, each night. Part of the purpose of this ceremony is to express gratitude for a good year, for rain and the abundant supply of plant-based foods and medicines. In certain ways it is like a fall harvest ceremony, celebrating the gift of surplus enough to share with family or trade with neighbors. At the end of the ceremony, the Holy Woman breaks her fast by eating a bowl of soup made from fresh serviceberries.

And the cycle will begin again.

Returning

Last summer my two adult daughters and I drove to the Big Snowy Mountains in central Montana. We wanted to return to places important to the Blackfeet that I had heard about from my grandmother and other elders, and from historical research. As Indigenous people it is difficult to set aside our history of colonialism, dispossession, deportation, displacement, genocide, and cultural genocide because our history is so much a part of our present-day lives. And to a lot of Indigenous peoples, "wilderness" lands are not beautiful natural landscapes but land that has gone feral due to the lack of human intervention. We tried our best to set all that aside, to immerse ourselves in the Blackfeet landscape, its mountains, its rivers, and especially its plants. Gazing across the steep rocky canyons and sloping prairies, we tried to envision these places not as US-defined wilderness (or wilderness study area) but as my grandparents saw them: as places cared for by our people, cultivated, developed, and managed for abundance for thousands of years.

Prairie Sonnet

Gwen Nell Westerman

In every season, we walk the prairie
Among the sustenance and the flora.
Shifting light and color creating awe,
Always ever-changing in its beauty.

Wild Rye waves in the wind as the Prairie
Turnip, Milkweed, and Coneflower there stand,
Yellow, pink, blue flowers thrive on dry land.
Still waters, a Lotus sanctuary.

Mulberries and Sand Cherries from the trees
Close to the prairie's edge defy death's maw.
Inspired by wind and unwritten law,
Red willow sways on ancient inland seas.

From beyond human knowledge, flora sing
Songs they carry for every living thing.

They Could Hear the Buffalo

Elsie M. DuBray

I think often about the beauty of the Buffalo on our homelands. The way their brown hues provide just enough contrast to catch your attention and yet blend into the landscape in a way that asserts they are meant to be here. They strike a delicate balance between unapologetic boldness and the tenderness of the land's dear friend.

I can think of nothing else when it is dusk in the summer and it's so still that among the birdsong and chirping crickets, I can hear the orange calf quicken his gait. His silhouette is dancing across the hot pink horizon, hastening to keep up with his mother, whose knowing grunts of reunion put me at ease, too.

I find that I cannot look away when it is February and the blizzard calms, and I can finally see the herd trudging along in the snow. Foggy breaths billow from behind the intricate white masks the winds have painted on their faces. They walk into the wind, facing the storm head on, and remind me that I, too, must go on.

My love for the Buffalo is older than anything else I could ever experience—a love from the cosmos, birthed in the Black Hills, nurtured and passed down by my ancestors. A love that courses through my body, physically, with every pump of blood through my heart. A love that tells the story of who I am and who I will become, from my people's creation, to how my parents met, to me writing here now—and everything in between.

This love grounds my commitment to a life of Buffalo restoration, the history of which in the so-called United States is fraught with contention, white savior–esque environmentalism, and omnipresent attempts of control from colonial institutions. Yet Native peoples have maintained our unique relationships to the Buffalo despite it all. Many Native-led efforts across Indian Country sustain this important work. As a Lakota woman whose creation story tells that we and the Buffalo are one and the same, it is with deep gratitude and admiration that I applaud those involved who have committed themselves to their peoples, to the Buffalo, and to the ongoing fight against settler colonial violence—including my parents and especially my father.

In the 1980s, my dad kicked off what would become the rest of his life's work and the foundation of mine. In 1990 he started working as a Tribal planner, tasked with Community and Economic Development initiatives for our Tribe (Cheyenne River Sioux Tribe). After going into every community on our reservation and speaking with people— listening to their needs, stories, and complaints—he approached Tribal council with the only conclusion he could draw from the overwhelming desire communicated to him: We needed to focus on Buffalo. He was eventually given the green light to build up our Tribe's Buffalo program, which he brought from 70 to 4,500 head while also managing a personal herd.

Witnessing the truly transformative power of Buffalo restoration on our reservation, he invited other Tribes who might also be interested in bringing Buffalo back to their communities to the Great Sioux Nation meeting in the Black Hills. With only three days' prior notice, nineteen Tribal representatives showed up to talk Buffalo—an unprecedented feat and testament to the cause. This incredible turnout and conversation led to the formulation of the Inter-Tribal Bison Cooperative (ITBC). My dad served as president of the board for several terms, while still working for our Tribal herd. He left the position with our Tribe and went on to become the executive director of ITBC until 2006, when he decided to focus on raising his children and managing our family's herd, which afforded me the upbringing that made me who I am.

It is the greatest blessing and privilege of my life to have grown up on my homelands surrounded by the Buffalo, picking wild plums and digging *tinpsila* (wild prairie turnips) for jelly and soup, and stopping to stuff my pockets with *ceyaka* (wild field mint) while playing down by the river with my brother. My backyard was the middle of a Buffalo herd! I hold immense gratitude for this life and do not take it for granted; my family's work and my dad's stories over bottomless bowls of Buffalo soup have fed my sense of purpose and stitched together the fabric of my being.

In my own life, I've carried the story forward through international science fair projects on Buffalo versus beef lipid profiles; food systems panels, presentations, and fellowships; an honors thesis about intergenerational Buffalo restoration and Lakota futurisms; Native community center jobs; an MS in community health and prevention research; and late-night roommate dreamscapes about the Buffalo and a world where freedom is true. The work I do, the thoughts I think, the compass that guides me, and the force that propels it all forward—I owe it all to the Buffalo and to my dad and everything they have taught each other and passed down to me.

While my degrees are not in botany or environmental science, I know there's a long list of chemical reactions and complex symbioses that a textbook or research paper could cite as to why Buffalo are essential to the restoration of our prairie ecosystems. But I don't want to write about carbon sequestration or the mechanisms through which Buffalo wallows contribute to micro-ecosystems and increased plant diversity on the plains. Because to me, that is but one small footnote in the larger story of my homelands, the story written into who I am.

It is one thing to appreciate the way that Buffalo enable and encourage perennial plants' return year after year and how those perennial plants contribute to the microbiotic soil health of the land. And I do. It's another thing, however, to appreciate that these plants hold within them the energy of my ancestors' bones and root their stories, carrying within their very cell walls an intimate understanding of my family, my people, my home, myself.

In my short twenty-some years as a Lakota, Nueta, and Hidatsa woman raised on the Cheyenne River Reservation, I have already witnessed and felt firsthand, like most of my community, the consequences of colonial violence. These consequences are not abstract for us. My family has endured the suicide of my eighteen-year-old brother Beau in 2020, my cousin six months prior, and countless classmates and community members. Funerals punctuate some of my earliest memories, and it seems a weekly occurrence to hear of another death back home. But the Buffalo have held me as I've reckoned with these losses.

I often think back to a specific day back home a few months after losing my brother. I was having a particularly hard day and feeling suffocated by the weight of all I was trying to make sense of. I felt moved to go outside. I went for a walk around home, finding a bit of relief in the evening air and solace in my dog Bella as we made our way out to visit a group of our horses. I was pleasantly distracted but soon became overwhelmed by it all again, breaking down into uncontrollable sobs. My horses heard everything I didn't have the words to say as I wept to them and cast desperate thoughts and prayers into the open prairie. I eventually surrendered to the ground as the sun set, tears flowing until they weren't, and I felt an almost unnerving sense of gratitude. Every second felt like an eternity as my home held me in one of the most vulnerable moments of my life and took some of my pain from me.

There is nowhere else in this world that could have tended to me the way my homelands did in that moment. Not just because it's where I grew up and have so many memories of my brother but because this land and these perennial plants know my grief and share it. When the juneberry bushes heard my sobs and the prairie grasses felt the dripping of my tears, I believe they remembered the stories of my ancestors that are tied to mine. I believe they remembered my brother and the time he spent laughing and crying and playing on those same grasses, picking those same juneberries, and I believe they missed him, too. I also believe these plants remembered the atrocities that my people, the Buffalo, and this land have endured. I think they made the same connections between that violence and my brother's suicide, and all the other premature deaths in my family and my community, because the stories written into our bones are the same absorbed by their roots. I believe these plants reconcile the same thoughts I do and pray the Buffalo come back, too.

Every day that I come closer to understanding the magnitude of knowledge my homelands hold, I feel more deeply the ways in which all colonial violence is connected, homelands across the world rooting down their own webs of the stories we share. Quill Christie-Peters (Anishinaabeg), in reference to Indigenous solidarity against the genocide of Palestinians, writes about the necessity of allowing oneself to rupture with rage and love and pain, positing that the act of breaking open allows us to become whole; it is a demonstration of our humanity and a rejection of normalizing settler colonial violence.

Losing my brother, I felt the undeniable power to make change that lies in the act of breaking open. The embodied experience of grief enables us to imagine, dream, create, and build an "otherwise world" grounded in our humanity. The Buffalo have taught me how to grieve and how, though painful, grief can generate more hope than despair.

The Buffalo teach me and my people how to return to them and to ourselves—and what it means to be a good relative. They carry themselves in a way that holds me accountable to a liberation we all deserve and demands that a true and just Buffalo restoration movement inextricably requires a free Palestine, Congo, Sudan. Buffalo practice a powerful form of resistance, persisting on this landscape despite so many attempts to eliminate them. They make the space within me to rupture, teaching me what Buffalo restoration means: to be whole.

That moment grieving my brother on my homelands was a turning point in my heartbreak. It deepened my relationship to the Buffalo and land and was the first time I felt empowered by grief's capacity to move me forward toward actualized hope. This is the same thread woven into the stories my dad tells of how Buffalo restoration on Cheyenne River brought healing to our community in real time. He saw our people—a people whose entire existence and creation story come down to a relationship with the Buffalo but who have had no day-to-day interaction with these animals for generations—reclaim the experience of watching them on our homelands as an everyday family pastime. To me, this is radical. The idea that on a regular day, Lakota people could stop to see the Buffalo as a normal part of their lives.

Recalling the early days of the Buffalo restoration movement, my dad said that the time he felt most hopeful was when a group of "elders from our community said they could hear the thunder of the Buffalo coming back and it was like music to their ears."

I can't hear them yet, but I can feel them. I feel them when I look at my dad, as I write their stories, and as I breathe, and I know the land feels them too. I look forward to the day when, alongside my community and the plants that remember, I can hear them, too.

There is strength in always returning, and it is my homelands that teach me this. So, like the Buffalo, like the plants of my homelands, I keep coming back. To these stories, to these teachings, to myself. Unapologetically. Perennially.

Brilliant Flames

Mariah Gladstone

On the Northern Plains and in the shadows of the Rockies, fire-deprived landscapes are apparent. You'll see large shrubby cinquefoil, sagebrush, and wild roses. Black angus cattle are speckled across the land, partitioned by barbed wire, along with the occasional mule deer or herd of antelope. The rolling foothills are nearly topographically identical to the past. But the composition of the animals, plants, and soils across these millions of acres has been dramatically altered with the erasure of Indigenous firecraft.

Growing up in Montana, we learned to associate summers with fire season and to associate fire with destruction. Dry summers meant the air would be thick with smoke. Roads were closed, and in the most extreme times, evacuation orders were issued and buildings went up in flames. Even as Blackfeet, we were not taught about our ancestral use of this tool and how we employed it to tend woodlands and prairies that fed our nation. Instead, we were given the Smokey Bear treatment and taught that all wildland fire is bad and should be stopped at all costs.

California was the first to criminalize Indigenous use of fire as a tool, with the 1850 "Act for the Government and Protection of Indians." When the Weeks Act was passed in 1911, traditional Indigenous burning practices were essentially outlawed nationwide. Similar laws were enacted across Canada, prohibiting Native people from intentionally setting landscape fires, while fire use by settler farms continued to be permitted. The settler state viewed trees as a commodity to be bought and sold as lumber; forests were resources to be mined, not lively supermarkets. Without careful tending by fire, our orchards were overgrown and our rich grasslands smothered with shrubs and aspen.

In my childhood on the Blackfeet Reservation, I learned how deeply land stewardship is woven into our culture, shaping both community and ecological balance. My education as an environmental scientist gave me the tools to understand how cultural fire, a practice Indigenous peoples have carried out for generations, is not only a vital part of our traditions but also a powerful method for restoring landscapes, fostering biodiversity, and building climate resilience. Cultural burns have long been used across the North American continent to encourage ecosystems of abundance that support diverse plant communities and megagrazers.

And, I came to understand, these cultural burns that were once the norm on this landscape were very different from the destructive fires that punctuated my childhood memories. Native people set these fires in the early spring or late fall when they could be closely regulated and would burn more slowly than fires that occur during hot, dry, summers. These cooler, low-to-the-ground fires cleared out woody growth to increase ecosystem productivity and create prime grazing land for herbivores.

In forest biomes, such low-intensity fires generally affect only the understory level, opening up pathways and allowing rapid regeneration of food plant species. Research from Elk Island National Park in Canada studied the effect of these fires on the growth of both native raspberry and beaked hazelnut plants, adjusting for fire severity. Although both species regrew the year after the burns, the lower-intensity fires showed higher numbers of hazel sprouts and increased raspberry height growth.

Similarly, other berry bushes including huckleberry, serviceberry, and thimbleberries thrive in a postburn environment. Fireweed, as the name implies, quickly returns to the charred landscape. Edible tubers such as camas and yampah (wild carrot) also bounce back readily after a low-severity fire, as the perennial part of the plant is not damaged by the fire. Because camas plants need sunlight to bloom, they won't thrive in a shady understory. Fungal spores also are safe underground, meaning mushrooms such as edible morels emerge quickly after a burn. In contrast, a high-severity fire, like the uncontrollable summer infernos I was taught to fear, can damage even the top layer of soil, taking longer to regenerate vegetation.

In northern California forests, Native people burned off forested areas in order to clear conifers and allow pyrophytic plant species to prosper. This included the thick barked oak trees, whose acorns were a critical food resource for tribes of the region. In Kentucky, researchers analyzed historical charcoal and pollen records to identify patterns of burning in the region. Unsurprisingly, charcoal deposits were high during the same periods when there were also high pollen levels from fire-adapted tree species such as oak, chestnut, and hickory. These tree species could survive low-intensity fires while benefiting from the nutrient recycling and additional sunlight the fires offered. As an act of reciprocity for the Indigenous caretakers, the trees offered edible acorns, chestnuts, and hickory nuts as rich food sources. Interestingly enough, the studies also found an abundance of pollen from sunflower and goosefoot, other plants with long histories of Native agriculture.

In grassland biomes, my Blackfeet ancestors and other Indigenous people of the plains used burns to manipulate the movement of herds. My elders have reminded me that a blackened prairie will warm faster, since the soil absorbs heat more than the surrounding old dried grasses. This triggers the germination of seeds as well as new grass growth earlier in the season. Bison and other grazing animals preferred these freshly burnt fields above rangeland. Actually, elk, antelope, bighorn sheep, mule deer, and white-tailed deer all prefer these burned areas. Research from the Jemez Mountains in New Mexico shows a historical correlation between herbivore numbers and increased burning. The charcoal data indicate that those anthropogenic fires were much less severe than modern wildfires in the same region and increased the food resources available.

What's more, most plant matter consumed by grazers is returned to the land in ways that other plant communities and microorganisms can use. The presence of large herbivores on a landscape ensures that at least 80 percent of forage nitrogen returns to the land, mostly through urine. Having these animals on the land provides a super-concentrated fertilizer to the plant and soil communities, boosting herbaceous growth. In addition, burning of the dung piles on these grasslands creates high-quality burn residue that feeds the soil phosphorus and potassium and further nourishes the land.

Certain plant species actually rely on prairie fires to assist in the germination of their seeds. For example, prairie turnips, large edible tubers that thrive in sandy soils, have a thick seed coat that prevents germination unless broken. Low-intensity cultural fires help ensure that seeds can germinate the following spring. Without it, the seeds may have to wait for several freeze–thaw cycles in order to break the seed coat and germinate.

It may seem like common sense, but dead leaves, grasses, and other biomass can quickly build up, causing an accumulation that deprives the surface of sunlight. This not only slows the growth of the plants but also makes it harder for herbivores to forage, causing them to exert more energy to get less food. It can take years for that leaf litter to decompose, but fire breaks it down into usable nutrients in minutes. Of course, the fire also reduces woody shrubs to ashes above the surface. The shrubs can regrow from their roots, but it takes several years for them to become sizeable. However, in the absence of fire, the shrubs continue to grow, gradually becoming more dominant.

Sometimes Native people used this fire-pruning technique to create intentional growth patterns. For example, Karuk people from northern California burned landscapes and then returned one to three years later to harvest beargrass leaves for use in traditional basketry. The same was true in places with willow. Burning off the aboveground stems would ensure that the next year's stems would be perfectly straight and ideal for weaving.

Today's grasslands look different from the prairie gardens tended for millennia before. The banning of Indigenous fire, slaughter of bison, and tilling of land to plant monocultural fields have taken a toll. The benefits of prairie fires for sustaining grassland ecosystems have been recognized, though, and fire has continued to be used for settler agriculture. It's often called prescribed burning, and it's driven by the desire to reduce deadfall in order to prevent catastrophic wildfires. However, these prescribed burns are disconnected from the Indigenous peoples who developed and fine-tuned methods of working with fire. While recognizing that total fire prevention has been devastating to the health of the landscape, prescribed burns also still position humans and nature at odds.

In contrast, cultural burning shows a deep, intertwined relationship between people and the land we steward. Cultural burning is still being taught and sometimes practiced across the United States and Canada, with a new generation of Native pyrohorticulturists being trained. By recognizing the empirical data collected by generations of Indigenous scientists, we can see ourselves as part of the ecosystems that surround us. We can tend these ecosystems as vast gardens.

As someone who works to revitalize Indigenous food systems and educates about traditional foods, I want to see both the return of Native ingredients and the return of land stewardship practices that promote the regeneration of biodiverse, food-producing grasslands and forests. Cultural burning is a critical practice, not simply as a defense against devastating fires but as a brilliant tool for creating vibrant forest gardens and prairies that can once again be harvested like grocery aisles. Today, I weave together my training as a scientist and the teachings of my elders to incorporate these ingredients in the kitchen as well as cultivate them for years to come. As a future ancestor, I'm responsible for restoring our stewardship to the land and the land's foods to our bellies.

Regeneration on the Range

Paige Stanley

Rangelands are unruly, which, if I'm being honest, is probably why I'm drawn to them. They resist even definition. Some are grasslands, while others are dominated by shrubs. Some are flat enough to see across the entire expanse, while others have mountains that loom over you like watchful giants. And some are owned by people in private homesteads, while others are held in a collective by governments for public access. Despite their differences, nearly all rangelands have at least two things in common. First, they are places that have been historically grazed by wild ruminants and are now predominantly grazed by livestock on native vegetation. Second, they refuse to be industrialized and commoditized—they are too dry, too rocky, or just too stubborn to be farmed or developed into something else.

I first found myself drawn to rangelands as a graduate student beginning a master's of animal science degree at Michigan State University. Many are surprised to know that I didn't come to love and study rangelands after growing up immersed in agriculture. Instead, I grew up in an entirely blue-collar family in rural Georgia and only stumbled across issues of agriculture and climate change during my second year as an undergraduate at a small liberal arts university. What began as a casual interest in reducing the greenhouse gas footprint of livestock production turned into an all-consuming curiosity. Could it really be that improving the way we graze livestock on rangelands could not only produce food in these expansive, unruly places but also sequester carbon in the soils underneath our feet . . . and hooves? Before I knew it, I was out on the range before sunrise, coffee in one hand and soil auger in the other, with dirt under my nails and wonder in my heart. In essence, I came to love rangelands and soil carbon by way of cows.

I've now spent the last decade of my life researching regenerative grazing systems and soil carbon sequestration. I've collected more soil samples than I could begin to count (about 8,000 samples is my best guess) on ranches of all sizes, histories, management strategies, and conditions. I've had my boots on the ground collecting those samples in every fathomable weather condition, from triple-digit temperatures during long droughts to sleeting rain so cold that it stings your hands. I've seen soils so dark and sludgy they resemble tar and samples so dry and coarse they slip through your fingers like beach sand. I've talked to ranchers and producers across kitchen tables, over Zoom, on the phone, on old truck bench seats, while loading bulls, and even over Facebook Messenger in some cases.

And what I've come to realize is that regenerative grazing is not a one-size-fits-all solution. Instead, it's a dynamic, context-dependent process shaped by the land, the people, and their goals. Its potential to address climate challenges lies not in its simplicity but in its complexity—intertwining age-old traditions with new science, human stewardship with animal instincts evolved over millennia. When we get it right, grazing strikes a delicate balance of ecological give and take, much like the wild herbivores who once called these rangelands home.

Over the thousands of years that bison freely roamed intact prairies prior to colonization, they moved in response to predators and ecological cues like forage quality and quantity, which were also shaped by Indigenous land stewardship. After grazing an area, bison would go on their merry way, traversing long distances in search of a fresh salad bar, giving the leftover plants time to rest and regrow. Over time, perennial grasses and plants evolved in tandem with this bison dance: They learned to grow deep roots and store nutrients to boost regrowth after they'd been grazed. In turn, bison learned when to graze plants to meet their nutritional needs and when to stop to ensure meals in future years. This coevolutionary relationship is what ultimately made North American rangelands highly effective at storing carbon in their soils. Bison—by simply eating grasses according to their genetic programming and signals from the environment—maintained rangelands as open spaces instead of forests. And grasses—by simply trying to survive being bison breakfast—learned how to stay alive and allocate more of their energy belowground, which grew the soil carbon bank account for thousands of years.

Today, most ranchers work with domestic cattle on smaller acreages than their historical and highly mobile bison analogs. But in these settings, many of the ranchers I've collaborated with in my research speak of trying to recreate ancient grazing patterns using management. They move animals frequently in response to ecological cues rather than allowing them to graze the same plants over and over again within a fenced pasture. Above all, they make these decisions adaptively: When their animals begin to look less healthy, or more bare soil starts to peek through the grass, or rainfall is less than projected, they change course. By trying to mimic and reconfigure what bison once did, these ranchers inadvertently cultivate a soil carbon frenzy right beneath their boots and spur a chain reaction that pulls carbon from the atmosphere, depositing it in small amounts belowground and helping to restore a long-lost cow–carbon–climate balance.

Making grazing management decisions based on ecosystem signals is no simple feat. To successfully adapt grazing to the myriad environmental cues of rangelands first requires ranchers to be able to recognize them, then internalize and interpret them, and lastly to understand the range of appropriate responses. Dogged determination, attention to detail, creativity, and ability to learn from one's mistakes—these are what I consider to be the attributes of the best scientists I've ever met. So I think of the regenerative ranchers I work with who exhibit these same attributes as nothing short of scientists themselves. Their grazing is a continuous experiment, one that I have the honor of documenting.

However, documenting regenerative grazing is a scientific challenge unlike many others. This quickly became apparent to me as a graduate student scouring Google Scholar, when I realized there were only a handful of studies to draw on. It turns out that the very nature of regenerative grazing, particularly the bit about adaptation, makes it both unrealistic to replicate in an experiment and impossible to control—eliminating two of the most fundamental tools for designing successful scientific research. Luckily, I like puzzles, and as an early career researcher, I was inclined to think a bit more out of the box.

On one of the first regenerative ranches I ever sampled, sunset came before we finished, so we resolved to complete our work in the morning. The next day, we came back to discover that all the flags we'd left in the pasture to mark our sampling points were gone. The rancher had decided that our sampling pasture was where he wanted to move his animals, rather than the pasture he'd originally planned, because the forage quality and quantity looked better. And it happens that little red flags are like magnets to curious, pesky cattle. While that made for a very long day for my team and me, I reflect on that experience now and realize that was when I finally came to understand just *how* adaptively regenerative grazers make decisions and the types of questions they ask themselves each and every day. These decision-making *processes*, and not just their outcomes, became a central focus of my research.

When I pull my first soil core on a new ranch, I'm giddy to see what it looks like. *What stories will it tell? What secrets have been waiting here underground? Is that black speck at 70 centimeters deep pyrogenic carbon from a fire hundreds of years ago? Are those soil aggregates in the surface soil the handiwork of sticky root exudates, and could that be the outcome of years of regenerative grazing practices? How long did it take that live root to get all the way to the bottom of my 1-meter soil core?* I've learned regenerative ranchers feel a similar curiosity every morning. *I wonder how my native perennials are recovering in pasture six. Did I leave my animals in that last pasture too long? Should I give the upland pasture more rest next year? What would happen if I left 60 percent of the forage in the next pasture instead of 50 percent?*

All of animal agriculture used to be more like this. Before chickens were bred to rapidly grow large breasts in industrial barns, before sows were confined to tiny crates during gestation, animal husbandry implied careful study and stewardship of the evolutionary relationship between animals and the plants they eat. This art has been almost completely lost in the hog and poultry sectors, as breeding has made it possible to raise animals in confinement on feed for their whole lives.

But cattle, like the rangelands where they originate, have a somewhat stubborn genome, so many of them still spend much of their young lives out on the range, foraging their own diet. Of course, most of these cattle are eventually headed to a feedlot, where industrialization and its social and environmental costs become all too apparent. But as regenerative ranchers recover the lost art of choreographing the dance between plants and ruminants, and some choose to raise animals on grass throughout their lives, I'm hopeful that we can create food production systems that work with natural processes instead of destroying them.

Whenever I need a reset, a reminder of why this work is worth it when we still have so far to go, I head back out to the range. All those data that have been waiting to be cleaned, coded, and analyzed? I'll get back to them first thing tomorrow. Just as soon as I watch the animals graze and marvel at how they know exactly where and when to take their next bite. I need to hear the rowdy sound of rattlesnake grass as it wakes up in the wind and watch how the shifting wild light changes purple needlegrass from deep violet to maroon to soft dusky rose.

One Prairie Strip at a Time

Lisa Schulte Moore

I bet I've seen more sunrises than you have. This is the one bet I'm willing to make with just about anyone.

As a scientist who studies the natural world and our impacts on it, I get up early. My circadian rhythm shifted in my twenties when summer jobs researching bird communities required 4 a.m. wakeups. The pattern stuck and now helps me balance the simultaneous demands of mothering and professoring. It also comes in handy when my current work on the sustainability of Midwestern agriculture calls for early mornings.

One dewy dawn in 2016, tromping through a field at daybreak to check on a vesper sparrow nest, is particularly memorable for me. It wasn't just any vesper sparrow nest but a specific nest in the middle of a 230-acre corn field in central Iowa. The corn was already tall and dense, such that I could only see a few feet in front of me.

I knew the nest had failed, but I had to make sure. As I tell each new cohort of student technicians, "Our credibility begins with high-quality data." According to the datasheet, a technician found the nest eleven days earlier, on June 15. An adult bird flew away upon approach, revealing four white eggs, less than an inch long each, with purply brown markings. From this information, I inferred that at the time the parents built the nest, the field would have been mostly barren, with rows of tiny corn plants poking through the soil. To them, the area may have looked somewhat like the short, sparse grassland that comprises a vesper sparrow's natural habitat. Another crew member checked it on June 19 and indicated things were still on track. The technician who checked on June 22 didn't see an adult, however, and the nest had just two eggs instead of four. By my visit, based on what's known about the life history of vesper sparrows, the eggs should have hatched, and both adults should have been tending to their babies. But the now tall, closed-canopy vegetation would not have been the home the would-be parents had bargained for.

When I reached the nest, it was as I expected: no parents, no eggs, and no young. The nest contents were probably eaten by a raccoon, coyote, or deer, and the parents had flown away in search of better habitat. What I did not expect, however, was the near-absence of life in my surroundings. The only two species I could sense were myself and the fast-growing corn. "I'm in a green desert," I thought. But that's not right: Though vegetation may be sparse, life is still abundant in nature's deserts.

The reason for my visit was to understand how returning bits of farm fields to prairie might restore the benefits that people derive from nature, today often referred to as ecosystem services. This field is a part of the Science-Based Trials of Rowcrops Integrated with Prairie Strips (STRIPS) project, established by my colleagues at Iowa State University along with partner institutions. We strive to understand and communicate the value of prairie strips: small patches of reconstructed, diverse, native grassland purposefully placed within annual crop fields. In an experimental trial starting in 2007, we found that replacing 10 percent of small agricultural catchments planted to corn and soybeans with prairie vegetation yielded dramatic environmental benefits with limited effect on crop production. These benefits include reducing water runoff by 37 percent, sediment loss by 95 percent, phosphorus loss by 77 percent, and nitrogen loss by 70 percent. We also found that prairie strips improved several soil health measures and habitat for pollinators and birds, including some species of conservation concern. Prairie strips provide these benefits because—unlike our annual crops—they are composed of diverse perennial plant species that together have abundant root systems, active throughout most of the year. Furthermore, unlike conservation cover typically established on farms, the stiff stems of many prairie species stand erect during driving rain, helping to slow down water runoff. Based on the results of our experimental trial, the STRIPS team took up the mantra of "small changes, big impacts!"

But could we replicate these encouraging findings on commercial farms? We were uncertain, and my quest with the nest in the large corn field—a "control" field in our expanded research program—was part of what we needed to do to find out. Adjacent to the control field is a "treatment" field, which is similar in every way except for one thing: It has prairie strips. The change is limited—10 percent of cropland seeded to prairie strips—but the differences are vast. On that morning, the prairie strip was teeming with life! Rather than just a single hue, I was greeted by a paint store's worth of greens, each associated with a different species of grass. I found a colorful bouquet of wildflowers dotted among the grasses: orange butterfly milkweed nestled between a trio of yellow species (gray-headed coneflower, evening primrose, and compass plant) while pale purple wild bergamot intermingled with the more intense blue vervain and the delicate white heads of wild quinine mingled with the spiky ones of rattlesnake master. Birds were singing, bees were buzzing, crickets were chirping, and butterflies flitted everywhere. The contrast made me feel like I was immersed in a work of art—the biological version of beholding a Van Gogh canvas while listening to a Beethoven symphony—situated among tidy and uniformly verdant rows of corn.

Over time, STRIPS team scientists have rigorously collected, analyzed, and published data, providing quantitative support for the differences I experienced that day. Fields with prairie strips support more individuals and species of wild bees than control fields without prairie strips; many of them show improved body condition. Honey bees raised on fields with prairie strips are healthier. Their hives produce more honey and are more likely to make it through the winter. Beetles also benefit: When comparing prairie strips to adjacent cropland, we found twenty-three more carabid species, a taxonomic group of beetles that can benefit farmers by eating weed seeds and the larvae of crop pests. There are twice as many birds on fields with prairie strips and nearly three times as many grassland birds, the group of birds experiencing the most precipitous decline in North America. And, as I was helping to document that June day, birds are more successful at raising their young on fields with prairie strips.

Important to our farmer and farmland owner partners, research on their commercial fields has by and large confirmed what we observed in our initial experiments, while expanding our understanding of how prairie strips perform under a range of cropland and prairie management conditions. For example, we now know that prairie strips especially improve water infiltration and thus reduce runoff in the fall and, potentially, early spring. Relatedly, runoff from crop fields with prairie strips contains 90 percent less total suspended sediment—also known as soil particles—and 90 percent less total nitrogen, 90 percent less total phosphorus, and 88 percent less dissolved phosphorus, compared with similarly farmed fields without prairie strips.

The soil metrics associated with prairie strips tend to be what initially captures attention. After experiencing a flood in 2019, farmer cooperator Jon Bakehouse decided he "needed to get more serious about our conservation practices. . . . If you're standing in your field and watching your soil, your resource base, wash into the ditch, it's hard not to take action after that." Our partners also want to be good stewards of water quality. Farmland owner Eric Hoien describes an epiphany while flying over the Gulf of Mexico: "This water quality stuff is real. There is a whole lot of Iowa in the soil you see down there, and that's not right. . . . I was wondering why anyone wouldn't be interested in keeping their soil and nutrients from moving."

Once prairie strips are established, people also notice the quick return of abundant life. Farmer Dick Sloan enjoys seeing flowers, pollinator insects, bald eagles, white-tailed deer, and ring-necked pheasants in his prairie strips. "The only bottom line is not the dollar bill," he says. "There's a lot of bottom lines for me that have to do with the environment." Farmland owner Shami Lucena Morse is excited about "improving the microbial community in the soil and providing habitat for wildlife—mammals, insects and birds." Many of our partners speak about enjoying the beauty of prairie strips and about how their family members are also attracted to the flowers. Stories of a granddaughter's senior portrait with prairie strips as the backdrop and prairie flowers on a stepson's wedding banquet table follow. Farmer Gary Guthrie wants "more wildness in my landscape . . . in the middle of this monoculture . . . for aesthetic beauty and for the native bees and the insects and the birds." Jon Bakehouse also talks about the mental health benefits prairie strips bring: "I have prairie strips so I have a good place to go to recover from a bad day."

Because of my scientific and personal experiences, I'm enthusiastic about the ability of prairie strips to dramatically improve ecosystem services on row crop agricultural landscapes. I also know prairie strips are neither a panacea nor an easy choice. Whole-field grassland reconstruction is a better option if the goal is to eliminate soil erosion or nutrient runoff completely. But these are working lands: Changes must be compatible with farmers' operations, and that's not always straight-forward. Establishing prairie strips can be challenging: Drought, torrential rains, pesticide residue carryover, weed pressure, overspray, and lack of resources for proper management can doom great intentions. Moreover, they take up space that could be in crop production. We've found that yield loss at the field level is proportional to the land placed in strips. Who, regardless of profession, wants to lose out on income? Given the high risks and often low margins associated with farming, I understand the reluctance. The fact that so many farmers and farmland owners still find a way should be a source of hope and optimism for us all!

My colleague Dr. J. Arbuckle, extension sociologist at Iowa State University, has investigated farmers' perspectives on prairie strips. The data he's collected indicate that Iowa farmers know about them today, which wasn't true when the STRIPS team started. Since awareness is the first step in behavioral change, this is really good news. Even more exciting: Over time, a greater proportion of farmers have indicated a willingness to plant them on their farmland, an increase from 14.8 percent in 2018 to 24.2 percent in 2024. Furthermore, 43 percent report an increase in their willingness to establish prairie strips if they receive remuneration by enrolling those acres in the US Conservation Reserve Program. These self-reported survey data parallel statistics from the US Department of Agriculture, which show a dramatic increase in land enrollment in CP43—the federal practice code for prairie strips—since it became available in December 2019. Today, there are more than 6,500 acres of prairie strips in Iowa, up from 388 acres before the availability of federal support. And it's not just Iowa: Interest is also growing nationally, with more than 26,000 acres planted across fourteen states.

With more than 250 million acres in row crop production in the United States according to US Department of Agriculture data, we have a very long way to go. Yet there appears to be widespread openness to prairie strips as a strategy to restore land and life in these fields. We have also seen that prairie strips can be a first step toward further regeneration and rewilding efforts: Several of our partners who started with strips are now on paths to convert their whole farms to grassland. While each farm operation presents a unique situation, it's clear from our partners' words and actions that these little ribbons of prairie can be a gateway to bigger change. In recognition that scaling up will be a marathon and not a sprint, the mantra of the STRIPS team is now "one prairie strip at a time."

The story of prairie strips illustrates one way in which US farmers and farmland owners can embrace our perennial roots—and the many benefits of doing so. We're still at a very early point in terms of cultivating broad acceptance for reintegrating natural and agricultural systems to support society's many needs, but I am hopeful for the future.

Are there prairie strips on a farm near you? If not already, I bet there will be soon. My tip is that they are best viewed at sunrise.

The Perennial Imagination

Jesse Nathan

The perennial imagination
nods in the bull thistle.
Lingers in the sap
at the sunlit needle's tip—
glimmers there like an earring.
Resists the wind. Is the wind.
Climbs barns with ivy
bent on bringing them down.
Works in the hands
that build back the barn.
Persuades berry to darken
and fatten, tickles
the branches, lets fall a windfall.
Listens in the gossip
of the grasses, dissolves
the clouds, dying on its back
with the junebug, faltering

the blind hound, trilling
in a mockbird, devouring
fire, obsessing worms, repairing
the shattered Judas Tree, lifting
water from an ancient bed, waking
in the root, scrawling
runes in frost on windowpanes.

From Corn Belt to Pasture

Laura Paine

For perennial agriculture to take hold, we need to address the elephant in the room: the fact that agriculture in the United States is, almost by definition, an annual affair. We'll never get anywhere with our quest for perennialization if we don't address the 90 million acres of annual corn grown each year and the fact that the agriculture economy is shaped around these commodities. The food industry, the agriculture input sector, even the financial investment sector live and breathe by this handful of annual crops.

I've worked in this male-dominated world of Midwestern agriculture most of my life. Over the last thirty years, I've had a lot of conversations where farmers eventually get around to asking, "Did you grow up on a farm?" For a long time, I've simply said, "No, next question?" I interpreted this classic back-and-forth as a test to see whether I measured up. They were asking whether I have any *real* experience or whether it was all just "book learning." But then I realized that maybe all these farmers were simply curious why someone like me would choose a profession like this. That's actually a good question. Why do I do the work I do, and why did my husband and I buy an 80-acre farm in our mid-forties? Actually, it all started in a field of annual corn.

It's 1971. The sweat trickles down my face, making rivulets in the dust and sticky yellow corn pollen that coats my skin. It's my first job, detasseling corn in a central Illinois seed corn field for $1.25 an hour. Each crew member takes a row, and we walk across the field pulling the male tassels out of the top of each corn plant before they can shed their pollen onto the female silks below. For every five or six rows of detasseled corn, there is a row that gets to keep its tassels, supplying the pollen for all and creating the hybrid between the two varieties.

My hometown of Decatur, Illinois, is the North American epicenter of corn growing and processing. Even though the paychecks that clothed and fed me came from that industry, I was raised a city kid. Detasseling was my first introduction to agriculture. It didn't scare me away. On the contrary, the head-high corn plants, the leaves rustling in the wind, the damp soil underfoot, and the sweet smell of corn sugar lit a little spark of interest. That was the start of my journey into agriculture as an occupation, but getting the credential of "being a farmer" would have to wait another thirty years.

Three years later, my future husband, Bill, and I are doing plant surveys in prairie remnants along railroad tracks as part of my first botany class at Southern Illinois University. In contrast to the sterile monoculture of the corn field, these railroad prairies are lavish. We run a 100-foot string in a straight line across the narrow prairie strip and count at least fifty different species in one transect. And the diversity doesn't end with plants. We flush meadowlarks and savanna sparrows from their nests and pass 3-foot-diameter webs strung between the waving grass stems, each occupied by a fist-sized yellow and black garden spider.

In what's now called the Corn Belt, railroad corridors were carved out of native prairie a century before the region earned its moniker. Until recently, railroad companies had limited tools or resources available to control plant growth along those corridors, and native plants flourished undisturbed. Not so with the rest of the Illinois landscape. The deep, rich soils created by tallgrass prairie rapidly became the center of grain production. A hundred years ago, the Prairie State already had nearly 9 million acres of corn. By 1974 when I did my surveys, prairie acreage in Illinois was down to 2,300 acres, less than 1 percent of its original area. The railroad prairies are linear gems of diversity, serving as an archive of the once ubiquitous grassland.

As I graduated from college, newly armed with a degree in botany, I began casting about for a meaningful occupation. Rachel Carson's *Silent Spring*, published in 1962, ushered in the environmental movement, pushing back against the prevailing Western paradigm of subduing nature. A new paradigm arose: one of protecting nature from seemingly inevitable human-caused destruction. Neither sat well with me. Couldn't we find a way to produce enough healthy food to feed humanity *and* protect the resources on which we all depend?

I've always wanted to believe the answer is yes, but it was another fifteen years before I stumbled across a way forward. The template for ecologically sound, profitable, regenerative agriculture, I realized, was a well-managed pasture. After all, *we* can't digest grass, but our faithful partners in agriculture can. Ruminant livestock have coevolved with humans since the dawn of agriculture. It has been a good working relationship. In exchange for a safe, well-fed life (probably longer than they would have experienced in the wild) and a guarantee of reproductive success, cattle historically provided work, milk, and meat—probably in that order of importance. Before the last seventy years of fossil fuel–driven mechanization, draft animals were essential partners of agriculturists for millennia. Today, work again may be one of the most important roles that cattle can and should play in American agriculture—the work of restoring the ecological benefits of perennial grassland to our agroecosystems.

The practice of managed grazing is patterned after the grassland ecosystem, allowing us to benefit from the synergies that make natural systems so efficient. Rather than maximizing yield or profit, we're looking to optimize the functioning of our farm ecosystem. Buying a farm at the age of forty-six was my way of trying my hand at making this theory a reality. Nature makes it look easy, but more than two decades later, we're still learning how to do it right.

Our farm is an 82-acre microcosm of the Eastern tallgrass prairie landscape. Surveyor James H. Mullett was in this part of Wisconsin in October and November of 1834 and recorded that this quarter section was a mosaic of wetlands, oak savannas, and prairie. Homesteaded in 1850 by one Thomas B. Haslam, this property was never as thoroughly farmed as the neighboring well-drained Arlington Prairie soils. The Crawfish River meanders through it sinuously, not straightened and ditched as it is upstream. The 1930s-era clay drain tiles have long since failed, allowing seeps and springs to dot the landscape. Putting it back into perennial grassland was an easy decision.

By raising grass-fed beef on perennial pastures, we not only create a means of generating income but also use the cattle to restore the health of the ecosystem. We rotate the herd through a series of small paddocks, giving the pasture plant community time to rest and regrow between each grazing event.

Nature's farming system creates a stable community of plants and animals that efficiently captures the sun's energy, cycles nutrients, and conserves water. Across the planet, she has created climax ecosystems that are exquisitely suited to the climate and soils of the location where they are found. Every niche is filled; every resource is used and recycled over and over again. Yes, corn is a grass, but a corn field is a far cry from a functioning ecosystem. The agriculture of the Corn Belt with its predominance of annual crops is an invitation to nature to fill in the gaps—the empty niches—and to move the system back toward a stable perennial ecosystem. The more we can shape our agroecosystems in nature's image, based on what nature wants to be, the less energy and resources we expend trying to stop the inevitable process of succession.

The grassland ecosystem shows us resilience can be realized through adaptability, diversity, and redundancy—multiple species with overlapping roles. We see this in our pastures. While these simplified, "human-made" ecosystems are populated mostly by nonnative cool-season grasses and clovers, they are diverse enough to be resilient to weather extremes. Our management has allowed remnant patches of prairie cordgrass and sedges, angelica and anemones, to reemerge and thrive. And nearly all the plant species in our pasture provide nutrition for our cattle—there's hardly anything you could call a weed. If it is hotter or colder or wetter or drier, different species dominate for a season, but the ground is always covered and there is always something there for the cattle to eat.

A grazing-based system creates complementary efficiencies for the farmer. If your cows harvest their own feed and spread their own manure most of the year, you can get by with smaller equipment, less of it, and fewer buildings—and you have less need for inputs like fertilizer and pesticides. This equation means that a grass farmer can steward her land, husband her cattle, and generate more profit per acre or per cow than the conventional livestock farmer. So in the end, a farmer can earn the same living on fewer acres with fewer cows.

Every Thanksgiving we make a pilgrimage back to central Illinois to visit family, a four-hour drive through the wasteland of bare soil that is Illinois in winter. On 22 million acres of former Illinois grassland, corn and soybeans are planted in May and ready to harvest by September, leaving the ground devoid of living plants nearly 70 percent of the time over the two-year rotation. In Illinois, 80 percent of the cropland is corn or soybeans; in the Dairy State where I live now, it's 43 percent. The difference is livestock, and the hay and pasture needed to feed them. If you think of this on a continental scale—over 90 million acres of annual corn nationwide—you can imagine how much ecosystem productivity is lost, how much organic matter storage is missed, and how many calories of the sun's energy are wasted, just warming bare ground.

Back home, my farm in early December is an island of green in a surrounding sea of bare, dead soil. While the neighbors' soils sleep fitfully through the winter, awaiting the plow and a new dose of fertilizer, seed, and pesticides, my soils teem with life. Although photosynthesis has slowed or stopped, diverse perennial pastures provide temperature moderation, consistent moisture, and diverse sources of nutrition that keep the soil microbiome healthy and thriving year-round. My soil is dark and crumbly like chocolate cake and has a sweet earthy smell, very different from the powdery gray soil in those cornfields I walked years ago. Annual tillage and synthetic fertilizers and pesticides disrupt the soil microbiome and burn soil organic matter.

In a perennial system, soil microbes—mainly fungi and bacteria—work in tandem with the plant community. Increasingly, researchers are learning how plants share the products of photosynthesis in exchange for microbial help in accessing minerals and water throughout the soil profile. It's a happy mutualism. Through these relationships, the reach of a plant's root system can be more than doubled, enhancing growth and allowing more energy to be captured, converted to sugars, and shared with partners. As soil organic matter accumulates, carbon stores grow and the functioning of the system only improves. But these relationships take time.

Recently, both the US Department of Agriculture and private companies have recognized that most annual cropland is dangerously low in organic matter and that agricultural soils can potentially serve as a carbon sink, helping address climate change. They have started paying farmers to adopt no-tillage systems and plant cover crops in hopes of building these soils back up. The problem is that it won't work.

The Wisconsin Integrated Cropping Systems Trial is a University of Wisconsin–Madison research project that has compared the performance of six cropping systems over more than three decades. All the annual cropping systems, including those using no-till and cover crops, have steadily lost soil carbon since the beginning. The forage-based rotations that include two to three years of alfalfa have also lost soil carbon. Only the rotationally grazed cool-season pastures and the restored prairie treatments have gained carbon in the top 12 inches. When measured across a 3-foot-deep soil profile, even these perennial treatments are only maintaining soil carbon levels. The notion that we can put back the carbon we've burned out of Midwestern soils is a false hope. Annual cropping systems will always be a carbon source, not a sink. Tweaking that system with no-till and cover crops is not a solution to climate change or a means of addressing water quality and habitat loss in the Corn Belt. Perenniality is the key to restoring the functions of prairie landscapes.

Maybe it's not realistic to shoot for restoring grassland to all or even most of those acres, but what if we think about a course correction? By bringing livestock back into Illinois' agriculture system, perhaps we can target a better balance between crops and forages, say 50 percent—roughly equivalent to the ratio in Wisconsin.

That would translate into about 11 million acres of pastures and hay and about 3 million head of cattle. That change has the potential to put 58 million tons of carbon back into the soil annually. Restoring those acres to perennial cover would also have a truly massive impact on water quality in the Mississippi River watershed. It would annually save an estimated 22 million tons of soil, 20 million pounds of phosphorus, and 217 million pounds of nitrate that currently wash into lakes and rivers and ultimately into the Gulf of Mexico. Eleven million acres of well-managed pastures would provide habitat for an estimated 28.6 million pairs of grassland birds.

But can we do it? Grass-fed products command a premium in the market, and demand is growing. Currently, more than 75 percent of grass-fed beef sold in the United States is imported from Australia and South America. Bringing that production home makes a whole lot of sense. It's a monumental challenge, and it's only gotten more daunting in the three decades I've been doing this. We've made some progress, and there's still a lot of work to do, but I can take satisfaction from knowing that I've found my path toward making a difference.

Out in our pasture, we watch as a calf comes into the world in a diving position: front hooves first with the muzzle close behind. It's wet and slick, sliding out ungracefully in a heap on the green grass, encased in the amniotic sac. Mom is on her feet within seconds, licking the calf all over, stimulating him to breathe. Within minutes he is wobbling around on shaky legs, getting his first meal of fresh milk. They are back with the herd within an hour or two when we move them to a new paddock. And within a few days, he is nibbling his first leaves of grass.

We have watched this miracle of birth hundreds of times on our farm. It never ceases to be awe-inspiring and humbling, a sense that is enhanced—not diminished—by this male calf's destiny as a meat animal. By raising the animals that sustain us, we become intimately engaged in their lives, immersing ourselves in the rhythms of life and death. Through them, we are connected to the earth's ecosystems in a tangible way that far too few people ever experience. With these animals and the grasslands that support them, we can help heal this hurting planet.

Root Foods

Kelly Kindscher

I live 30 miles downstream from Topeka, Kansas, a place named for digging prairie roots for food. In truth, *Topeka* is *dopeka*, literally "good place to dig *do*" in the language of the Kaw people who are native to this place. *Do*, known in English as hopniss or groundnut, thrived on the sandy soils of the Kansas River floodplain, where this plant could be readily harvested. But today, you'd be hard pressed to run into a Topekan who's heard of this plant by any name.

Few people know any prairie root foods these days, as they have been overlooked, plowed up, and grazed out. And now many of them are hard to find, rarer even than native prairies. Who knows tipsin, the prairie turnip? Who knows eastern camas, also called wild hyacinth? Who knows blue funnel lily or any of the twenty prairie plant root foods we have identified in ethnobotanical studies of the Prairie Bioregion?

My work as an ethnobotanist has its origins on a homesteaded farm near Guide Rock, Nebraska, where I spent part of my childhood. My great-great-grandparents settled there in 1871, only a few years after the Pawnee were forced to give up their large territory in Nebraska and Kansas. In this vast open country just north of the Kansas border, my dad taught me about pasture plants and encouraged my budding interest in the natural world. Imagine my disappointment when, as a senior researching an honors thesis at the University of Kansas, I came to realize that the homesteaded pasture on my family's farm was severely degraded. The land had lost many prairie species completely, and continuous annual grazing by cattle had greatly reduced numbers of the remaining plants. I wanted to know more about the plants that seemed to have gone missing from the land, their botany and ecology, and ultimately their uses for food and medicine.

I got more involved in the conservation movement, and I did a deep dive into edible plants, reading everything I could find. I was particularly intrigued when I learned about a plant known as *tipsinla* in Lakota—shortened to *tipsin* by Euro-American ethnobotanist Melvin Gilmore—which had once been plentiful in the very place where my family farmed. Then one summer as I was working on the farm, my mom and I joined a workshop with wild food author Kay Young at the Cather Memorial Prairie in Red Cloud, Nebraska. Young dug a tipsin plant, showed us the white and starchy root, and told the amazing story of this food below us. As their seeds ripen in July, all of the above-ground parts of the plant dry up, break off, and blow away, stem and all, which keeps them hidden.

Most prairie species are perennials, and many have deep rooting structures that enable them to find moisture and nutrients well below the soil surface, beyond the reach of annual species. A few of them also have underground storage organs. These storage organs are helpful to the plants, as they provide energy and moisture to ride out long droughts. They also provide food for animals, sustaining gophers, other rodents, and even the now-extinct Great Plains grizzly bear. And as the Kaw place name of Topeka indicates, these carbohydrate-rich storage organs can be a great food for humans too. But being a great food means the plants need to protect themselves from being eaten into oblivion.

So that's why tipsin hide as tumbleweeds! I was astounded to learn that this bean family species was once the most widely used wild plant food among the Native nations of the Prairie Bioregion. High in protein, with edible carbohydrates, this quintessential prairie plant features the turnip-shaped root Kay Young revealed to me. These root foods were cooked, eaten raw, or braided by their longer, stringy taproots and dried (similar to garlic braids) for later use.

It soon became clear to me that to learn about useful plants of the prairie, I needed to learn from the people who were here long before my ancestors. I continued to read the historical literature, to better understand the long-standing relationships with plants that have shaped Indigenous lifeways in this place. But I knew that there were also many contemporary Native knowledge holders who were still here. On the suggestion of colleagues who worked at Haskell Indian Nations University, near where I lived in Lawrence, I traveled to the Rosebud Sioux Reservation in South Dakota in hopes of learning from such knowledge holders. I ended up at the Ring Thunder community, where I met Alex Little Soldier, known more widely as Alex Lunderman in his eventual role as tribal chairman. Alex taught me about edible and medicinal plants, and even more valuable lessons about culture and respect and spiritual practices and ceremony. And I learned that Lakota still harvest their *tipsinla*.

Through my journeys and studies, I have learned about how traditional tipsin patches were known to Native people. When Tribal nations went out to the High Plains to hunt buffalo in the summer, camps were made near tipsin and chokecherry patches so that the women could harvest them while the men hunted bison. These tipsin patches were tended and harvested sustainably every few years. Mothers told their children that tipsin had families too and that their leaves pointed to each other. Then they encouraged their kids to go find them. This allowed the women to use their digging sticks in the hard ground without their kids in the way, and the game of finding the next plant was fun for the kids.

Timing was critical for this harvest. Native women waited to dig these plants until the seeds were ripe, as they wanted to collect the roots when they were as large as possible. But they couldn't wait too long. These women knew that the plants needed to be harvested in July, before their tops broke off and blew away.

To give thanks for the food, harvesters often returned the dried tipsin top along with some tobacco or corn meal, depositing these offerings in the hole dug when removing the plant. These tops were known to have ripe seeds too. This practice encouraged seeds to germinate in disturbed soil, where a new plant could successfully grow. In this way, healthy tipsin patches were sustained and could be returned to again in a few years once the plants had developed larger roots.

Now, when I think about the prairies I know and love, and how tipsin plants have become less common in these places, I think about the generations of Native harvesters who have been prevented from tending the land and following their traditional practices. And I reflect on the communities who have persisted in tending tipsin and enjoying the nutritious rewards. People have the power to help tipsin populations and be helped in turn.

~

Tipsin is one of many prairie plants that have underground storage organs. My grassland travels to Topeka and beyond have also led me to hopniss, the town's namesake. This root, sometimes known as groundnut, was probably on par with tipsin as one of the most important foods of the region before the introduction of corn. Nutrient dense and widely available, hopniss may have been even more central to prairie diets than wild game.

Hopniss is the Delaware name for the plant, which was used extensively by myriad Native communities, from the eastern part of the continent reaching west to the wet prairie meadows, ravines, and floodplains of the Great Plains. Topeka isn't the only spot on the map that evinces this legacy. Place names in Native languages for several rivers flowing through present-day Nebraska, Kansas, and Oklahoma bear witness to hopniss habitat and the importance of this plant. Unfortunately, few stands of hopniss remain, as the floodplain prairies where they were once harvested have been tilled for agricultural use. Grazing has reduced or eliminated hopniss as well, as livestock enjoy this plant's tasty, tender, protein-rich foliage. But there's still some out there if you know where to look. I have found hopniss in the moist, sandy soil of ungrazed creek bottom pasturelands in central Kansas, with strings of egg-shaped roots connected underground, seemingly ready to harvest.

Research at Louisiana State University in the 1980s offered some new hope for hopniss as a future food crop, as horticulturalist Bill Blackmon selected for larger tubers, obtaining yields of over 8 pounds of tubers per plant. Today, there is yet again interest in hopniss as an alternative crop with significant potential. Japanese farmers and researchers have been developing this root food as a specialty crop, as they have a similar native species. It is not surprising that hopniss would be taken into cultivation in wider environments, as this root crop is high in both carbohydrates and protein, with surprisingly high amounts of vitamin C.

But to unlock all this nutrition after harvest, people need to know how to prepare root foods. Many of these plants produce bitter compounds or store carbohydrates in less digestible forms as another means of "hiding" from herbivores, so they can't necessarily be eaten raw. One of the key strategies used by Native peoples to improve the digestibility of root foods is pit-baking or hot rock cookery. The technique is perhaps best known in the Pacific Northwest, where many Indigenous communities share a practice of gathering camas bulbs and slow baking them for days until they become sweet. This technique was also used on the prairie, as we now know from evidence at archaeological sites from Texas to Kansas. Pit baking was useful because the long period of heat exposure converted a greater percentage of the starchy and nondigestible inulin (the major form of carbohydrate in camas and other bulbs) to sugar.

It's clear that root foods have been a central source of nutrition in this part of the world for hundreds of years. Some bulb remains found at the Camp Bowie archaeological site in Central Texas date back to 1,500 years ago. But I would argue that they may be even more important in the future, as we all reckon with climate change and the need to adapt the ways in which we grow food. This climate future will require food plants that are more drought tolerant and resilient to ups and downs in the weather, exactly the traits that these prairie root foods have evolved. We'll also need to find more ways to harvest our meals from diverse perennial ecosystems that function like the landscapes that were here before agriculture. Root foods can help us restore the prairie while providing nutrient dense meals. I'm well aware that folks don't necessarily think of Kansas as a culinary destination these days, but maybe its fertile soils will once again be celebrated as a good place to dig dinner.

Prairie Root Foods

Species	Common Name	Root Type
*Allium canadense**	Wild onion	Bulbs
Amphicarpaea bracteata	Hog peanut	Tubers
Androstephium caeruleum	Blue funnel lily	Bulbs
Apios americana	Hopniss or groundnut	Tubers
*Callirhoe involucrata**	Purple poppy mallow	Thickened roots
Calochortus gunnisonii	Sego lily	Bulbs
*Camassia scilloides**	Wild hyacinth	Bulbs
Claytonia virginica	Spring beauty	Tubers
Cucurbita foetidissima	Buffalo gourd	Thickened roots
Cymopterus acaulis	Stemless Indian parsley	Thickened roots
Erythronium mesochoreum	White dog's tooth violet	Bulbs
Glycyrrhiza lepidota	Wild licorice	Thickened roots
Helianthus tuberosus	Jerusalem artichoke	Tubers
Hoffmannseggia glauca	Indian rushpea	Tubers
Ipomoea leptophylla	Bush morning glory	Thickened roots
Liatris punctata	Gayfeather	Thickened roots
Lilium philadelphicum	Wood lily	Bulbs
Lomatium foeniculaceum	Prairie parsley	Roots
*Nothoscordum bivalve**	False garlic	Bulbs
Pediomelum esculentum	Tipsin or prairie turnip	Thickened roots

*Denotes species found in pit-baked archaeological sites.

The Beauty of Polycultures

Valentín Picasso

My elementary school textbook taught me that "the richness of Uruguay lies in livestock grazing our native grasslands." And I whole-heartedly agree that the native grasslands—"Campos" in my country and "Pampas" in Argentina—are a true treasure. Warm season grasses dominate the Campos, while cool season grasses dominate the Pampas. Along with a wide array of plants, a diverse fauna of mammals and birds thrives there, including *carpinchos* (capybara), *mulitas* (small armadillos), *zorrillos* (skunks), *ñandúes* (ostrich-like birds), and *dragones* (beautiful yellow-black birds).

"Grown-up" people unfortunately cannot understand impor-tant things without numbers, as Saint-Exupéry explains in *The Little Prince*. So, for grown-up readers, this biome provides habitat to 4,000 native plant species, 300 species of birds, 29 species of mammals, 49 species of reptiles, and 35 species of amphibians. These grasslands also provide multiple ecosystem services: They cycle nutrients, protect soil from erosion, and store carbon, which helps mitigate climate change. From this ecosystem, pure water flows to creeks, streams, and rivers, such as the Uruguay River. Even the name *Uruguay*, meaning "river of the painted colorful birds" in the native Guarani language, evokes biodiversity.

I have fond childhood memories of visiting my family's sheep and beef cattle ranch near Cerro Chato ("flat hill") in the easternmost corner of the country. I enjoyed riding horses and had lots of fun running behind the sheep in the corral while my dad cleaned and cut hooves and counted heads (numbers *are* important). But my favorite pastime was to walk through the Campos, finding different grasses and flowers, marveling at unusual rocks and critters. And I loved getting lost in the short, dense riverside forests by the banks of Las Pavas ("the hen turkeys" stream) and discovering ferns, lichens, and fruits of native trees.

While studying agronomy in university, I collected different native grasses from the Campos and tried, not always successfully, to identify the species. The grass family became my favorite botanical group, and *Paspalum dilatatum* (dallisgrass or honey grass) was my favorite species. *Stipas*, *Boutelouas*, *Axonopus*, *Stenotaphrum*, and *Bromus* also held a place in my heart. The more I got to know these grasses, the more concerned I became about their gradual disappearance from the landscape.

The Rio de la Plata grasslands that comprise the Campos and Pampas are one of the most threatened biomes in the world. Some of these grasslands have been overgrazed, and others have been replaced by monocultures of annual crops like soybeans or forestry plantations of eucalyptus and pine trees. If I wanted to contribute to solving this problem, I realized that I needed to understand the big picture, with all the messy social and environmental elements of agricultural systems.

So, after graduating with my "Ingeniero Agrónomo" degree, I joined the Graduate Program in Sustainable Agriculture at Iowa State University, in the heart of industrial agriculture in the United States. I wanted to better understand what fuels the mindset that more production is always better. And I was hoping to find answers about how to conserve biodiversity, soils, and water while still producing food and livelihoods for people.

My first week in the United States, as part of my first graduate course, the visionary professors I was lucky to find took us on a road trip. Among other stops, we drove 403 miles (648 km) to Salina, Kansas, to visit The Land Institute, then a tiny nonprofit organization, consisting of a house, an old barn, a greenhouse, and a small classroom. Its president was Wes Jackson, a university genetics professor who left academia and co-founded the organization to search for a more sustainable agriculture.

Wes explained to me and my fellow students that there are many problems *in* modern agriculture: soil erosion, water pollution, biodiversity loss, fossil fuel dependence, poor profitability, and the disappearance of rural communities, to name a few. But the real problem *of* agriculture is that we are growing annual monocultures that in no way resemble natural ecosystems like prairies or grasslands. Natural ecosystems have two important features: diversity and perenniality. They foster many different plant species that regrow every year. A truly sustainable agriculture, one that regenerates itself, must mimic the natural ecosystems in those two aspects. Furthermore, in Wes's words, "humans are grass seed eaters." Nearly every civilization was based on domesticating some annual grass: wheat, barley, rice, maize. It seems that I was not that unique in my predilection for the grass family! So The Land Institute's goal was to develop perennial grain crops and then to grow them in diverse mixtures or polycultures.

This encounter changed my life. The call for more perennials and more diversity in agriculture made so much sense to me. I got it. My native Campos are already sustainable, I realized. They are not only a beautiful treasure. They are a model for agricultural systems! Ever since then, my research has aimed to understand and design diverse and perennial agricultural systems, from grasslands to perennial polycultures.

After obtaining my PhD, I returned to Uruguay as a faculty member in my hometown university. I was happy to be back among family and friends, enjoying our tasty *asados* (the grass-fed beef, prepared over a wood fire, that is a signature of Uruguayan cuisine). And I joined a team of researchers studying the sustainability and resilience of livestock systems in our native grasslands. We demonstrated that well-managed grazing farms promote biodiversity and ecosystem services while providing healthy food and livelihoods. Just a simple practice like moving the animals to a different paddock when the grass sward height reached a minimum, we found, could double production of forage and animal weight. Optimal grazing management can preserve grassland diversity while increasing meat productivity in these agroecosystems.

And then came the threat and reality of climate change. For a country like Uruguay, with many more head of livestock than people, measuring and reducing the carbon footprint of livestock production became a national imperative. The president of the National Meat Institute (yes, Uruguay has one!) asked me whether it was true that cows were causing climate change, and the Minister of Agriculture and Livestock was also concerned. A famous report from the United Nations Food and Agriculture Organization had estimated that "livestock's long shadow" contributed more to global climate change than the transportation sector. Although that claim was not quite correct, the contribution of livestock to climate change was real. In Uruguay, we had no scientific assessment of this assertion; we needed grown-up numbers. So we started to quantify the carbon footprint and other environmental impacts of the country's livestock production. Our team compared three different ways of raising cows: on native grasslands, on sown pastures, and in feedlots.

Wait, feedlots? You might ask why a country that teaches its elementary school students that naturally available forage is its greatest treasure would resort to such a thing. Ironically, the rise of feedlots was the result of well-intentioned efforts to *reduce* the environmental impact of agriculture. In the 1990s, Uruguayan farmers rapidly adopted no-till methods, in an effort to curb soil erosion. By the end of the decade, the practice had been established on 90 percent of the country's agricultural land. But this technology also made it possible to grow crops in places where tillage was not possible because of their shallow or rocky soils. These seemingly "marginal" areas are some of Uruguay's most biodiverse native grasslands. About 1 million hectares of grassland was converted to agriculture between 2000 and 2015, changing the country's livestock industry. Instead of finishing beef cattle on highly productive sown pastures, surplus grains from agriculture were fed in feedlots.

Our early research results seemed to suggest that finishing livestock in feedlots might reduce emissions. Cows gain weight more slowly on pasture than they do in a feedlot, giving them more time to belch methane (the potent greenhouse gas that results from their digestion) into the atmosphere. However, feedlots have three negative environmental impacts that outweigh their seeming efficiency: They increase soil erosion, contaminate water with nutrients and pesticides, and result in more fossil fuel use. We also had to account for the fact that converting grasslands to agriculture is a major contributor to climate change through the release of soil carbon. When we factored that conversion into our models, we learned how grazing systems could potentially help recapture some carbon into soil. Overall, we found that optimally managed beef production based on native grasslands is more sustainable than grain-based feedlot systems. The message got to our policymakers, and programs to protect native grasslands as adaptation to climate change were strengthened.

Today I work far away from my native Campos grasslands. However, that beautiful agroecosystem is still my model for the diverse perennial systems I am researching. In my lab at the University of Wisconsin, we are designing new agricultural systems that cover the soil year-round, produce perennial grains for food and forage for livestock, and nurture deep roots that protect the soil, store carbon, and maintain clean water. As I walk with my students through the campus research farms at the end of harvest season, I point out the contrast between our green perennial polycultures and the brown of the surrounding annual crop farms. I tell them to keep an eye out for the birds that will visit as our experiments burst forth into many shades of green in the spring and blossom with colorful flowers in the summer.

One of those birds, the bobolink, is a familiar friend from the Campos. Like me, it's journeyed here from South America, where it spends the winter. In recent decades, bobolink populations have declined sharply, as intensive agriculture has encroached on their habitat. So seeing one here always makes me smile. Perhaps this bird, so content in our perennial research plots on a summer day, might have spent the previous January in one of those vast plains I roamed as a child, the national treasures I worked so hard to conserve in my early faculty days in Uruguay. In human terms, these two places are often considered wholly separate. A grassland in Uruguay. An agricultural field in Wisconsin. But in bobolink terms, they are two points along what might one day again be a great grassland corridor, bustling with life.

III. GRAIN

GRAINS ARE SEEDS WE EAT. Once ripened, these seeds come loose or can be threshed from the rest of the plant and harvested in mass. Hard, dry grains can be conveniently stored, traded, and transported. From the plant's perspective, seeds are the promise of a new generation, and they contain all the energy needed to help sprout a new life. So when people consume seeds as grains, we are accessing that valuable stored energy.

Most people alive right now depend on grains for essential carbohydrates, proteins, and fats. We especially depend on just a few high-yielding crops: Rice, maize (corn), and wheat together provide more than half the calories humans eat directly, mostly in refined forms. In less affluent countries, the direct reliance on grains for calories is even higher. (Through industrial food systems, people also increasingly grow grains as commodities to feed to nonhuman animals which we use for meat, eggs, and dairy. Factoring in this indirect consumption of grains, they account for 70 percent or more of the calories humans consume!)

We enjoy cereal grains in bowls of cooked plump rice, kernels of corn ground and shaped into round tortillas, and wheat flour kneaded into dough and baked into bread. More cereals are oats, barley, rye, and sorghum.

The grains of legumes are known as pulse crops. Pulses include lentils, chickpeas, and beans of many colors, shapes, and sizes. Pulses are second only to cereals as the most frequently consumed grains in the world, often in protein-packed stews or fiber-filled dips paired with rice, pita, or roti.

And oilseeds, like sunflower and canola, are critical for cooking and used in packaged goods. More than any other crop category, oil crops have increased in global production since 1961, due to both higher yields and larger harvest areas.

Grains are the hearty base of many food cultures. But major grain crops and all the people who depend on them are vulnerable, challenged by stagnating yields and lack of diversity. Our current agricultures are not yet the more sustainable, resilient, nutritious, or equitable food systems we need to help navigate climate change.

That's in large part because we rely on just a handful of staple crops, and all of these are annuals. Biologically and ecologically, the world's major grain crops come from short-lived plants with fairly shallow roots. Annual grains are planted on more than two thirds of the world's arable landscapes.

Cultivating annual grains requires energy to till the soil for planting and weeding each year and to harvest, process, and distribute crops. Historically, that energy came from the labor of humans and other animals; now, that energy comes especially from fossil fuels, which power machines and produce fertilizers and other agrochemical inputs.

Healthy forests, prairies, and other terrestrial ecosystems don't need such intensive inputs to be biologically productive and provide ecological benefits. Moreover, these diverse perennial ecosystems support the fundamental processes that sustain life and land, from water and nutrient cycling to soil protection and formation. Right now, grains are not widely grown in perennial cropping systems, but people are currently working to develop crops from perennial (rather than annual) grasses and forbs. Based on what we know about the robust ecological services provided by grasslands, and early research results, we have strong reason to expect that perennial grains will restore ecological functions while producing staple foods.

In the twenty-first century, perennial grains are an increasingly possible reality. People are developing novel perennial grains through two main pathways, using old and new knowledge and tools with a keen sense of urgency.

Plant breeders are working to make crosses between existing annual grain crops and their wild perennial relatives, in order to bring the life strategy of perenniality into important high-yielding staple crops such as wheat and sorghum. This is the pathway being successfully pursued with the development of perennial rice, which results from hybridizing the annual domesticated Asian rice *Oryza sativa* with the perennial African *Oryza longistaminata*.

The second pathway, crop domestication, builds on the history of humans and plants coevolving to depend on each other. Humans learn to grow, tend, and value a particular wild plant species for food while selecting between those plant populations for the characteristics they want: bigger and tastier seeds, easier to harvest or thresh, more fit to human management in places like fields, more resistant to pests and diseases. These cycles of selection lead plant populations to become crops that feature those desirable traits. Perennial grain crop domestication is an ongoing process in which wild, already deeply rooted perennial plants are becoming grain crops. This is the pathway being taken with Kernza® perennial grain, harvested from selected cycles of the perennial Eurasian intermediate wheatgrass *Thinopyrum intermedium*.

Humans have long eaten the seeds of many wild grasses, including both annuals and perennials. And in Australia, Indigenous communities have harvested and processed perennial grass seeds as grains.

So why did some of our human ancestors focus on domesticating *annual* plants into what have become our current global grain crops? We don't know for sure. Breeding involves choosing and reseeding plants with desirable traits and removing plants with less desirable traits, to allow for crossing and selection. It might have simply been easier to work with shorter-lived annuals, which already died on their own, and to leave longer-lived perennials be (and to continue gathering from them) rather than remove most of them to facilitate selection. Or it might have been geographic circumstance: Although most vegetation is perennial, annuals yielding lots of seeds commonly find their place in naturally disturbed landscapes like the dynamic river valleys and floodplains where key human agricultures emerged.

What we do understand is the need to evolve our food systems now in the face of climate change and the challenging social, cultural, political, and economic systems from within which we must work. Human systems of domination have emerged and grown over the last dozen millennia with annual grain agricultures. The ways we relate to each other and the more-than-human world, and the ways we make a living, put some people above the rest of life in ways that undermine our collective ability to flourish.

Perennial grain agricultures could open up the chance for more just, mutually beneficial systems of relationship. This won't happen automatically; we'll have to work intentionally toward care, repair, and meeting shared needs. We will need to grow perennial grains at broader scales and in diverse systems that produce ecological benefits and delicious, nutritious foods. And we'll need to orient toward what is possible with less fossil energy but more creativity: resilient, perennial ways of being that can keep on evolving for the long term. As we work toward these ambitious visions, the people who are developing and stewarding perennial grains, and the places where these grains are valued and eaten, have much to teach us.

Australian Native Grains Awakening

Jacob Birch

I've always had lucid, dream-filled sleep, but not vivid. The dreams rapidly dissipate upon waking, like water soaking into the earth and evaporating after a summer sun shower. Human faces morph like reflections rippling in water. Words are tacit and coalesce as concepts. I take it as the calamity of the external world filtering through my internal psyche and not something on which to think too deeply. But when an ancestor came to me in a dream, it was profound. He appeared strikingly clear against an empty void, and he spoke one word in an ancient tongue that resonated through me like a tuning fork. I didn't know what it meant at the time, but I feel like it may have something to do with the path I have been walking since.

The path has led me to work tirelessly to revitalize an Indigenous Australian foodway and bring it into a modern context with authenticity and integrity. This foodway belongs to my people and was curated and passed on to us by our ancestors since the beginning of

time. We are Gamilaraay, people of the meandering rivers and grassland plains of southwest Queensland and northwest New South Wales.
Being grassland people, we have grassland foodways. The food I work
for, and with, is our long-lived perennial grains. Ganalay and Guli are
the names of two of our most important grasses. We use the gift of
their grains just as you would use wheat. The grains are small, but not
as small as amaranth or chia, and slightly resemble barley in color and
form.

I believe my ancestor visiting ignited this path of work for me. It
wasn't a conscious decision—I had no idea what the work would be
and what it would entail. But it felt like a calling home. My partner
and I were in Sardinia staying in a seaside town called Bosa. Our abode
was a terrace house carved into rock on a hillside that overlooked the
Mediterranean. At the top of the hill sat a castle that once housed
Crusaders. We were coming to the end of a ten-month journey through
Europe. It was meant to be three Australians, us two and my younger
brother, moving to London to work. But when we tired of that after
two weeks, it became a spiritual journey and a deep dive into permaculture and ancient European foodways, which stretched from Wales to
Norway, Moscow to Sardinia, and many places in between.

Upon reflection, Sardinia was pivotal. I remember the weathered red
earth and the perenniality of things. The ancient olive groves where
some trees had been around longer than Christianity. Down in the
valleys you would see a vibrant and abundant living mosaic: heritage
crops of wheat, barley, and rye; vineyards, vegetable gardens, orchards,
and olive groves; pigs, sheep, cattle, and horses. On the hillsides more
olive groves, interspersed with myriad other crops, such as almond,
and wild asparagus ready to be plucked for lunchtime pasta dishes. On
mountain plateaus were ancient cork oak forests where locals hunted
wild game and harvested mushrooms, as their ancestors had done for

millennia. Things weren't perfect. Many people were emigrating for urban lifestyles, and industrial farming was banging at the gate. But Sardinians were clearly hanging on to something that we in Australia had yet to fully reclaim.

Coming home to Australia I went down a few dead ends as I tried to navigate the path. There was the labor hire stint, when I didn't know whether I would have work one day to the next, and a traineeship that paid so poorly we couldn't afford baby diapers at one stage. But eventually, my partner encouraged me into university, and I haven't looked back. It was during my double degree in environmental science and marine science where I learned about the rapid and jaw-dropping degradation of land in Australia. We have already forfeited millions of acres of arable land to salinity, erosion, acidity, and the toll of urban sprawl, and we are forecast to squander millions more within the next few decades. We are losing land that could be producing food. We are losing land that supports our unique biodiversity. We are losing land that once stored millions of tons of carbon. We are losing land that sustained Indigenous people for thousands of generations. And we are losing it all at an incredible speed after mismanaging the continent since the First Fleet landed in 1788. In that short time, the damage that has been done is, soberingly, irreversible. Many of our plant, animal, and microorganism kin that helped sustain the complex web of life on our lands are now extinct.

In researching a solution for degraded landscapes, I thought, couldn't we just plant native, salt-tolerant grasses into salinity-affected land? In Australia, our approach has always been to replace native grasslands with introduced, highly invasive, "improved" pasture grasses, like buffelgrass. Among the charming qualities of buffelgrass is allelopathy, meaning it releases chemicals that suppress all other species, thus helping it form a monoculture. It does not provide any benefit to biodiversity, and it changes fire regimes, generating more dead biomass that burns more intensely. Cattle don't even like it. In contrast, our native grass Ganalay is a deep-rooted perennial. She is highly tolerant of our droughts, floods, fires, and nutrient poor soils. She produces high-quality fodder for cattle. And she is a keystone species, providing the habitat that supports everything from soil microbiology, invertebrates, reptiles, birds, mammals, other grasses, and herbs, right up to our totemic species, kangaroo and emu. There was only one link missing: Could our native grasses support human food security?

While searching the literature, I came across scientific articles mentioning native perennial grasses and their use as food for Indigenous people, my ancestors. I asked Elders in my family, and they nonchalantly confirmed this nearly forgotten part of our heritage. It was almost as if they were asking, "What did you think we were doing out here for the last 2,500 generations?" I guess you don't know what you don't know. That was the lightbulb moment, the hook that reeled me in and shattered colonial notions of Indigenous people as impotent and ignorant nomads living on the edge of existence, opportunistically spearing a kangaroo here and wandering across a berry there. The management of grasslands, the harvesting of grain, the processing of that grain, and the making of bread requires a stable, organized, and sophisticated culture—particularly in Australia, the driest inhabited continent on Earth.

Before colonization, Indigenous people meticulously curated the Australian landscape to ensure the provision of an abundant and diverse assemblage of food, fiber, and medicine. The worldviews, knowledge systems, ethical frameworks, and practices that were developed by the multitude of distinct peoples were equally vibrant and diverse. These peoples had a high level of autonomy yet adhered to strict sets of rules, which were governed by the lore and law of each nation. Furthermore, nations did not act in isolation from one another but instead knew all things are in relationship, being deeply interconnected and interdependent. Colonization has done a great job of dividing us.

The true history of colonization is not taught in Australia. We teach that Australia's identity was forged in the fires of World War I, where our legendary ANZACs commanded respect from other nations for their display of tenacity, sacrifice, courage, and mateship. We embrace this romanticized vision of ourselves. But we don't teach that while our young ANZACs were dying on the shores of Gallipoli, pastoralists and police officers were massacring Indigenous people back home. The Indigenous experience of colonization was disease, dispossession, the loss of traditional food systems, frontier war and indiscriminate massacre, internment on missions (our version of the reservation), separation of families and removal of children (our version of residential schools), and exploitation as slave labor. Many of these atrocities, which amount to genocide, were carried out in the name of agricultural expansion, particularly the pastoral industry.

The pastoral industry's impact on Indigenous people in Australia cannot be overstated. As early as the 1810s, Australia's colonial administration was promoting large-scale sheep operations that were akin to American open range ranching. In 1824, with significant financial investment from some of the wealthiest, most privileged, most well-connected men of the British Empire, the Australian Agricultural Company (AACo) was founded. The land grab by pastoralists looking to make their fortunes proceeded with a fervor generally reserved for a gold rush. By the 1860s, almost every patch of land on this continent where grass grew had been claimed, leaving only the harsh interior. To this day, AACo is still majority owned by a British multibillionaire and holds a land portfolio in Australia that is greater in size than the entire state of West Virginia. Outside interests bent our agricultural system for their own individual wealth creation, replacing the "culture" in agriculture with "business."

I saw the impacts of export-driven agribusiness when I worked several seasons as a "chipper," removing weeds from the vast cotton fields that now replace our perennial grasslands on Gamilaraay country. I spent my days traversing eerie moonscapes where only cotton grew. Human-made lakes stretched to the horizon, full of water siphoned out of our rivers. And I watched rural communities atrophy and suffer with chronic disease as wealth and decision-making power were exported to shareholders and boardrooms.

So, coming from the historical and contemporary context of Australia, Sardinia was a revelation. Sardinia is one of the world's blue zones, where people live longer lives than the rest of us. It's little wonder why, with such abundant, diverse, vibrant food systems. But it's not just the food that enables long life, it's the sense of community, it's the culture—a culture that predates the Romans.

Around the globe, we celebrate the cultural foodways of peoples whose regional gastronomies are intrinsically tied to their very identity: Champagne that comes only from the Champagne region in France, Scotch whisky that comes only from Scotland. Imagine the foodways of the world's oldest living culture and what gastronomic treasures they might hold. That is what I envision for Australia: not so much the export opportunity of a high-value niche product but localized, bioregional food economies that celebrate the unique foods and cultural heritage of Indigenous Australia, the world's longest-lived continuous culture. For Gamilaraay people, our region would become renowned for the foods, fibers, and medicines that symbolize our deep heritage and timeless landscapes.

Imagine touring through Gamilaraay country and visiting our perennial native grain production systems. You would see an old-growth forest in miniature, but more diverse, with grasses in place of trees. Like matriarchs of the plains, these ancient grasses that live for decades would gather an array of other beings into their embrace. Looking down into the center of a tussock, you might see a delicate legume, fixing nitrogen as a gift to Ganalay in reciprocal gratitude. Lifting your gaze to the horizon, you might see emus grazing the Queensland bluegrass that grows between the Ganalay tussocks. Imagine smelling the sweet, clear air while savoring a breakfast of native grain bread made of the unique assemblage of edible grains harvested from our perennial grasslands and cooked in our earthen ovens. And washing it all down with a fortifying drink made of a sweet tree exudate.

But when I imagine such a journey through Gamilaraay country, it doesn't end with food. I see a future where we would be known for our artifacts, carved from some of the hardest timbers on the planet. I see a sustainable timber milling industry supplying our termite-resistant timbers to carpenters. I see people from all walks of life coming for healing from our medicines and healers. I envision a society that makes holistic and integrated decisions that deliver social, environmental, cultural, and economic outcomes, for the benefit of all people.

So, for me, native grains present an auspicious opportunity. Yes, native grains contribute to carbon sequestration, biodiversity restoration, food security, food sovereignty, healthy nutrition, better resilience in the face of climate change, job creation, a just transition in a decarbonizing economy—and they can do all of this good work with minimal external inputs. But, more than that, the reawakening of native grain foodways can be an act of Indigenous self-determination and nation-building. The entire industry can be scaffolded around Indigenous values, ethics, knowledge systems, and worldviews. We might begin to remedy the wounds of the past and move toward a future where we all might thrive on this arid continent for another 60,000 years.

As I keep moving forward, I reflect on that ancestor who came to me and ask, "What is this work for?" I wonder whether that message was even for me to understand at all. This work will require an intergenerational effort, and therefore it will require a message that speaks to the next generations. Knowledge from the past to inform the future. So for me, that's the message: As descendants, are we doing the work to honor our ancestors? And when we are gone, will we be ancestors worth honoring?

Where the Story Starts

Muhammet Şakiroğlu

I grew up in eastern Turkey, in a highland village surrounded by mountain ranges. Covered with snow in winter months, these highlands
bloom in the spring with a massive variety of herbaceous plants. During
long-ago ice ages, the Black Sea, Caspian Sea, and Caucasus mountain
range protected these high plains from freezing and created a refuge for
plant species. Plants from all over Eurasia were pushed to Anatolia by
glaciers and forced to live together, which created the striking biodiversity seen across the region today. Earlier inhabitants of Anatolia took
advantage of this plant diversity and domesticated a large number of
crops. These crops include well-known annual grains such as wheat and
oats, plus many perennials including forage legumes and tree crops.
These plants were subsequently acquired by farmers across the world,
shaping the trajectory of global agriculture.

My family inherited a medieval watermill situated in the center of
this colorful meadow. As a gathering place for locals, the watermill
was where people passed on practical information about agriculture, as
well as folklore and gossip. Although some important crops had been

abandoned by the time I was born, watermill stories kept them alive, ensuring this farming knowledge was not lost to future generations.

One plant in particular stands out among my childhood memories, bursting forth in spring patches of unimaginably vibrant pink against the otherwise green shortgrass prairie of Ağrı (the easternmost province of Turkey). So pink, in fact, that the image in my mind almost seems borrowed from the colorful fantasy movies of contemporary digital cinema technology, with the glorious Mount Ararat in the background. As I rode the little bus that traversed the river basin from our town to the province center, these pink fields would suddenly appear, grabbing my attention. I would press my face against the window to catch a glimpse, attempting to pair the vision before me with the stories from the watermill. Eventually I found the answer: These were fields of *koringe*, as the Kurdish-speaking farmers in my community knew it. Or, in English, the perennial legume known as sainfoin.

I learned that sainfoin was planted for two purposes: providing feed (largely protein) for livestock and rehabilitating soil used for cultivating annual grains. The little communities scattered in the deep valleys around the rivers in Eastern Anatolia had relatively sustainable farming for many years because they cultivated a number of perennial legumes like sainfoin in their crop rotations. These perennial legumes hosted bacteria in their roots that could capture atmospheric nitrogen and transform it into an available form. This fixed nitrogen would then be deposited for the benefit of other plants growing in the same soil. These perennial legumes also served another purpose: feed for the horses and oxen that provided animal-based power. The animal-based power, in turn, was used in farming, construction, transportation, and many other human activities. Perennial legumes could almost be regarded as a sustainable form of fuel.

With the rise of fossil fuels and synthetic fertilizers, however, sainfoin and its cousins were gradually abandoned. For nearly twenty years, as I grew up and became an adult, I didn't see a single pink field. Giving up perennial legumes had two very dramatic outcomes for agriculture in the region. The first consequence was a long-term dependency on synthetic fertilizers. As farmers were forced to spend larger and larger portions of their incomes on imported fertilizers, their well-being began to suffer. Meanwhile, only some of the fertilizer was taken up by crops, while the rest ended up running off the field, contaminating creeks and rivers.

The second consequence of abandoning perennial legumes was a protein shortage for people. In order to produce meat, eggs, or dairy, farmers need access to enough plant-based protein to feed their livestock. When perennial legumes diminished, the number of animals in rangelands and small households decreased too, or these animals were moved into confinements that use imported protein like soy. As a result, a significant amount of the protein needed by the animal husbandry and aquaculture sectors in Turkey and the wider Middle East now comes from vast soy monocultures in the United States, Brazil, and Argentina. Large-scale transportation of annual pulses from production centers in the Americas to the Middle East requires an immense amount of fossil fuels, to say nothing of the unsustainable annual farming practices used to grow these industrial feed crops. In addition to the environmental costs associated with this food system, the financial cost of animal products is now quite high for the majority of the Turkish population.

The fascinating link between the fate of human societies and the plants around them became apparent to me while I was studying biology in college and reflecting on the lost pink fields of my youth. This fascination with plants' role in shaping both the history and future of humanity led me to grad school in botany and ultimately to a career in the breeding and genetics of cultivated plants.

Perhaps not surprisingly, I ended up working on sainfoin. As I considered how best to reintroduce perennial crops in the region, I went through a list of perennial forage legumes such as alfalfa, clovers, and bird's-foot trefoil. I had worked on alfalfa extensively and knew many of the other forages well from my study and research. But among them all, sainfoin emerged as the most promising. It performs well in nonirrigated conditions and is drought resistant. It does not cause bloat in animals and has good nutritional benefits. All in all, an ideal protein for livestock to graze on.

Meanwhile, on the other side of the Atlantic Ocean, another group of researchers had also started thinking about sainfoin, but as a human food. A longtime friend, Dr. Valentin Picasso, introduced me to Dr. Brandon Schlautman from The Land Institute, who was launching the effort. Their aim was to breed a new perennial pulse crop whose grain could be used for both animals *and* humans as a source of protein. Sainfoin seemed like a great place to start because it has relatively large seeds, similar to green lentils. Excited by this new prospect for sainfoin, I joined on as a collaborator.

Making a perennial pulse crop out of a species that has historically been used as a forage crop raises a lot of questions and challenges. Because forage crops are harvested for hay and then fed to livestock, their breeders have focused on enhancing the feed quality of the entire plant. However, converting a forage to a pulse changes the goal of breeding: The new focus is harvesting large amounts of high-quality protein from the seeds. Previously, scientists knew very little about the seed content in sainfoin because seeds have been used only to plant the next generation. So our team had to start with basic questions: Are the seeds edible? Do they contain harmful compounds for human and animal health? How do the seeds taste? Are they easy to cook?

So far, we've answered the most fundamental questions. Yes, sainfoin seeds are edible with no apparent problems. They contain a good combination of nutrients such as protein, oils, fibers, and carbohydrates, as well as beneficial minerals. We now also know that sainfoin seeds do not contain hazardous materials for human or animal consumption. We now call these seeds Baki™ beans; in Turkish, Baki means "long-lasting."

The next goal is to breed sainfoin to provide a high enough seed yield so that it becomes profitable for farmers and thus a viable alternative to current annual pulses. The real question is how long will it take to accomplish this. Our ancestors domesticated many different annual crops over a time period of ten millennia, or even more. Do we need to breed so many years to have a decent yield?

Fortunately, no, we do not need to wait so long for this crop to reach its potential. In the past, people selected seeds from their best plants for sowing the next year's crops. This traditional process of artificial selection has gifted us the grains we eat today. However, in the twenty-first century we now have a more robust knowledge of genetics that can help us speed up the process of selection. For example, my students and I spent many years focused on better understanding the genome of sainfoin and its plant relatives and helping build databases for sharing this knowledge. If societies invest in the technological and community resources that are needed, we could more quickly apply our scientific knowledge to develop perennial sainfoin as a brand new pulse crop.

Being one of the last of its kind, our watermill sat on the creek and served local farmers until the early 2000s. Regrettably, it was abandoned with the tragic loss of my older brother and the broader changes to agriculture. But not everything was lost.

Back in the watermill's heyday, I sat in the shade of the large willow tree next to it, memorizing sagas and stories about the history and farming of our village. I have passed many of these on to my kids while putting them to sleep. Some of my stories are filled with sainfoin: Heroes graze their horses, locals feed their livestock in the pink fields. In some stories the sainfoin is absent, as people abandoned old ways of farming and farmers forgot how important sainfoin was. But there is another story that I want to be passed from generation to generation. It is a new one: the story of the perennial pulse crop sainfoin. This is a complicated saga, about a legume that is still in the process of being rediscovered and reformatted as a pulse grain, making its way back to where it started. I do not have the whole story here, just a short beginning. The hero of the story is not one person but a group of devoted and talented people. The story is going to be told not only around a remote watermill in Kurdish but in books in many languages, shared not only with my children and grandchildren but with generations to come. I am hoping the story will be completed, whether or not I am around to read it.

Toward Perennial Sorghum

Pheonah Nabukalu

In the early morning light, perennial sorghum stands tall. The narrow, blade-like leaves glisten with dew, bending gracefully from sturdy, cane-like stems. The smell of damp earth fills the air, and as the sun rises, its warmth begins to chase away the coolness of the night. This simple scene is a reminder of the remarkable relationship we share with the land, one that has driven my work as a plant breeder.

My journey into perennial crops started with an unexpected fascination: sea oats (*Uniola paniculata*). During my graduate studies, I worked on breeding this subtropical coastal species, a grass that plays a critical role in stabilizing dunes and protecting shorelines. I spent hours walking along the beach, feeling the sting of salty air on my skin and listening to the waves crashing nearby. The resilience of these plants, which could withstand the harshest conditions, captivated me. The way they held the dunes together, providing protection from storms, made me realize how vital perennial plants are for maintaining ecosystems.

When I transitioned from coastal ecosystems to agricultural land-scapes, the parallels were clear. Just as sea oats anchor sandy shores, perennial crops like sorghum can anchor sustainable food systems.

I remember my first encounter with perennial sorghum. Shortly after beginning my postdoctoral fellowship with Dr. Stan Cox, I found myself standing amid a test plot in Kansas. The sorghum plants rose above me, their heads towering 8–12 feet overhead. As I knelt to inspect the soil, small shoots emerging from the earth hinted at what lay beneath. Digging deeper, I uncovered thick, cream-colored rhizomes: underground stems that store energy for the plant's regrowth. Holding them, I marveled at their role in sustaining life year after year. This moment ignited my fascination with this tenacious crop.

My journey with sorghum, however, began long before that moment. While I was growing up in Uganda, sorghum was woven into my everyday life. I remember watching my family and neighbors tend to plots crowned with clusters of small, round grains that ripened to a rich, earthy brown. The grains, with their mild, nutty flavor, were ground into flour for porridge or brewed into traditional beverages for gatherings. In times of drought, sorghum was a lifeline, nourishing us when other crops might fail. I recall my mother cooking thick porridge over an open fire and the warmth and familiar taste of this food bringing comfort. For us, sorghum was more than sustenance. It was a shared heritage, a link to generations past, and a source of strength.

Although the sorghum I grew up with was an annual crop, some traditional landraces could produce a second harvest, albeit a smaller one. This practice was called ratooning: Farmers would cut back most of the plant but leave the roots and just a piece of the shoots, from which a new plant would sometimes grow. When I learned that researchers were working on perennial sorghum—a crop that could return year after year with sustained yields—it felt revolutionary. This new crop could offer not only increased stability but also the promise of reducing reliance on resources like water, labor, and the cost of purchasing seeds each season.

Breeding perennial sorghum is no small task. We begin by crossing annual sorghum (*Sorghum bicolor*) with its wild relative, Johnsongrass (*Sorghum halepense*), in hopes of combining the best traits of both plants. Annual sorghum has the advantage of being nutritious and versatile, which is why it is already a major food source in parts of Africa and Asia. Our goal is to enhance its perenniality, making it more resilient to stress and more efficient in using resources like water and nutrients. This means selecting parents with the best attributes, crossing them, and then planting their offspring in different environments—from the humid fields of Georgia to the colder climates of Kansas to the tropics of Uganda. We closely monitor these plants, observing their growth patterns, flowering times, stress tolerance, and yield. In temperate climates, we track fall rhizome development and spring regrowth, while in the tropics, we assess their response to the return of seasonal rains. Every generation brings us closer to a crop that can thrive in different environments.

Developing perennial sorghum means working with the people who will grow and benefit from it. I often think about a farmer I met in Pallisa district in Uganda. His name was Opeto, and like many farmers in the region, he had struggled with increasingly unpredictable rains, which made farming more difficult. During my visit to his farm, he showed me a test plot of annual sorghum, his face lighting up as he pointed to one variety in particular. "This one, it will do," he said, nodding approvingly. "It stands strong when the others wilt and has better yield." Opeto's words stayed with me. They reminded me that farmers are not just passive recipients of agricultural innovation—they are partners in shaping the future of our food systems.

As I make return visits to Pallisa district, I'm excited to introduce test versions of perennial sorghum to farmers like Opeto. Their insights—on traits such as drought tolerance, disease resistance, and yield consistency—are crucial in ensuring that the crop is adapted to the specific challenges they face. We've embraced a farmer-centered design approach in which farmer feedback directly guides our breeding objectives, ensuring that the crops we develop are truly suited to their needs. Watching farmers evaluate test plots has been one of the most rewarding aspects of my work.

Of course, there are challenges. Some of the plants we develop still exhibit wild characteristics like unsynchronized flowering, low pollen production, or seeds that are difficult to thresh. There's also the risk that a perennial crop could become invasive if it's too successful at establishing itself. We have to strike a balance by creating a crop that's sturdy enough to survive multiple seasons without overrunning the landscape. As Andrew Paterson, director of the Plant Genome Mapping Laboratory at the University of Georgia, once told me, "We want enough perenniality to overwinter but not take over."

One of the biggest challenges we've faced is overcoming the ploidy barrier, meaning the difference in chromosome sets between Johnsongrass and sorghum. Johnsongrass is a tetraploid, meaning it has four sets of chromosomes, whereas most of the annual sorghum we use in breeding is diploid, with two sets of chromosomes. This difference makes crossing the two plants difficult. But in 2018, during one of our study trials, we observed plants that looked unusually perennial yet displayed many traits similar to cultivated annual sorghum: strong stems, large seeds, and a robust growth habit. These were our first diploid perennial sorghum crosses.

This success has accelerated the development of perennial sorghum varieties that could bring long-term benefits to farmers and the environment. It has also made it easier to study the genetics of perenniality and apply molecular tools like marker-assisted selection and genomic selection—techniques that are much more challenging to use with tetraploids. These technologies allow us to pinpoint the best genetic traits early in the process, speeding up the breeding work. It's like having a roadmap that shows us which plants are most likely to thrive. By integrating these tools into our program, we're making breeding not only faster but also more precise and effective.

At the end of the day, my work is about applying scientific knowledge to help create a more sustainable future. In regions like Africa, where sorghum is a dietary staple, developing a crop that can withstand drought and poor soils could support food security. It could also play a role in mitigating climate change by reducing the need for annual tillage, which disturbs the soil and releases carbon into the atmosphere.

As I walk through my research plot, I feel a sense of awe. Each plant holds so much potential, stretching far beyond the fields in front of me. I don't know exactly what discoveries today will bring, but I know that each step brings us closer to perennial sorghum.

Joyful, Needed Work

Lee DeHaan

When I was hired by The Land Institute in 2001 to help develop perennial grain crops, our work received little attention. At the time, perennial grains were the obscure dream of a small group of scientists and philosophers, and even they had varied opinions on the likelihood of success. Initially, our humble team worked quietly, searching among thousands of plants for unique individuals with bigger seeds and higher yields. When I noticed rapid progress domesticating a wild grass with the clunky moniker of intermediate wheatgrass, we started to test it in foods under the snappier name of Kernza. Farmers began experimental plantings, and word of the new grain began to spread. Journalists who took note of our progress didn't just want to know *what* I was doing but *why.* I prepared for interviews by summarizing my motivations: "Perennial grains like Kernza have potential to simultaneously produce abundant food for humans, provide environmental services, and improve farmer incomes."

Although my one-sentence answer explains why Kernza as a perennial grain is a good thing, it sidesteps the question as to why it is the one good thing I've been working on for over twenty years, out of so many possibilities. If I'm asked about my personal motivations, my mind always goes to the widely and variously quoted line from theologian Frederick Buechner: "Your calling is where your deep joy meets the world's deep need."

Buechner's famous quote has been debated by theologians and blog posters alike. It is tempting to think that following your joy might imply doing what pays the best, receives the highest praise, or keeps you in air conditioning on a hot day. Instead, I believe Buechner had in mind a richer understanding. When I was a young college graduate, my "deep joy" wasn't something that I could have clearly articulated. But in retrospect, I can now trace the process of discovering joy in my vocation.

I grew up on a farm and attended a small private school with five children in my grade. We didn't have many opportunities for sport competitions, but once every two years we had the chance to participate in a track and field competition with other schools. While my older siblings had brought home collections of ribbons in earlier years, I wasn't gifted with the ability or even much desire to excel in physical competitions. So not surprisingly, I was delivered home by the school bus on a warm spring afternoon without a single ribbon to show. Looking to take my mind off the day, I walked down the road to where my mother was driving a tractor to till the land for spring planting. I climbed into the tractor and rode along to visit and observe.

My wise and insightful mother, on hearing of the day's failures, decided it was time to take action. I had been copiloting tractors since I was a young lad, and even though I was only about twelve years old, I had long since learned all that was needed to operate the machine. So she turned to me and asked, "You want to drive the tractor?" I shrugged "I guess so." She got me started and left me on my own to work a field for the first time.

That afternoon I found joy. Undoubtedly like many boys I was thrilled by the ability to control a massively powerful machine, leaving churned soil in its wake, but the profound satisfaction I felt was something more meaningful. My work of the afternoon was producing food.

Although my family found fulfilment in growing food, the early 1980s were difficult years to be in farming. Right after we followed the advice of the experts and purchased more land and equipment, interest rates skyrocketed, and commodity prices plummeted in response to overproduction and weak exports. Land prices fell by 60 percent, meaning that loans for recently purchased farm ground were now greater than the value of the land. No matter how hard farmers worked or how brilliantly they innovated, external forces drove far too many of them to bankruptcy and even suicide.

The crisis in agriculture took its toll. My mother took a job in town. Dad was taking classes to become a real estate agent, preparing for a career after farming. Conversations among folk in the community were full of ominous words like "foreclosure." The often-repeated advice for youngsters was "Don't go into farming." Ultimately, we had to sell our family farm to a wealthy California investor for the cost of the debt. From that point on, my father became a farm manager, which was a surprising blessing that allowed our lives on the farm to return to something like normal.

Living through the hardships of the 1980s made me want to get away from farming. However, my father advised me to stay in the field of agriculture, even if I didn't become a farmer. He said that growing up on a farm, which he called "a farm background," gives you the sort of practical experience in agriculture that is hard to come by any other way.

My eldest brother was the first to put Dad's advice into action. After college, he spent three years in Nigeria, working in agriculture and missions. Once when he was back home on the farm, he said something that lodged firmly in my young teen brain. He said, essentially, that what the world needs is new crop plants. Corn and soybeans are widespread in Nigeria today, causing many of the same problems that haunt crop production in the Midwestern United States. My brother was going to graduate school, hoping to help create new perennial crop production systems.

Meanwhile, as I prepared to go to college, I considered which box to check indicating my intended major. Life on our farm bordering a Minnesota lake was close to the natural world. We watched migrating waterfowl from a canoe, hiked the forest to greet the spring wildflowers, and listened to the gentle sounds of nocturnal animals while camping. Threats to these treasures were being talked about in classrooms and on the news, leading me to choose environmental studies.

The first semester, I took Environmental Studies 101, which I found could have been more accurately named Depressing Topics 101. It is hard to find much to smile about when presented with one disaster after another, with most of the problems seeming insurmountable. At some point I remember writing to my parents that I thought I was coming to understand the biblical concept of "Blessed are those who mourn," as learning about the environmental crises we face produced a deep and unrelenting grief.

If the state of the planet with its contaminated waters, degraded soils, and astounding rate of species extinctions brings grief, I came to hope that doing something to shift course could be a path to joy. As I searched for my place within this vast array of environmental challenges, my dad's words came back to me. Maybe this was where my background as a "Minnesota farm kid" might come in handy. In summer evenings on the farm as the sun sank below the horizon, I often found myself walking through our vegetable garden. Daily observations of the slow, steady transformations that unfold as a garden grows brought a calm that I kept seeking out. Ten years later, I followed my brother to the same graduate program at the University of Minnesota and even inherited a box of seeds that he had collected from the wild that I used to initiate my PhD research on perennial legumes.

What I came to realize is that perennial grains are the amazing intersection of my deepest joys with some of the world's deepest needs. Here was a single change to agriculture that could enhance food production, improve farmer incomes through reduced expenses, and address environmental problems ranging from water pollution to climate change. And in that realization, I have come to see another joy, working to improve the overall system of agriculture rather than focusing on individual problems. Often our solution to one problem makes another problem worse. For instance, many techniques to address environmental issues in farming result in reduced yields or reduced farmer incomes. Perennial grains are so compelling to me because they offer a means to improve the system as a whole.

Perennial grains spoke to my budding concerns about the environment: Here was a solution that could reduce the burden of chemical fertilizers and heavy tillage while building up greater resilience in the soil. At the same time, perennial grains spoke to my inner twelve-year-old who loved growing food—and who also worried about his family's financial future. If farmers didn't have to pay for all those fertilizers, maybe they could make a small farm pencil out. I loved the idea of this virtuous cycle: farmers substituting ecological processes for industrial inputs, thereby realizing greater profits and better soil quality, which in turn would help revive rural economies and communities like the one I grew up in.

Of course, it's not quite that easy. Plant breeding takes years of dedicated work, and that work requires resources. So far, perennial grain breeding has yet to attract the level of research investment that would qualify in my mind as giving it a serious try. My guess is that the total historical investment in all perennial grain crops combined is still less than what is spent studying one of the major annual grain crops in a single year. And yet my colleagues in this field have managed some astonishing progress. After an investment of only about $20 million over fifteen years, researchers in China have developed perennial rice varieties that yield similarly to annual rice for eight consecutive harvests. These exciting results provide hope that many more projects, including my own with Kernza, might not be that far away from achieving breakthrough success.

Thirty-five years on from the conversation with my brother that first sparked my interest in perennial grains, the need for this work is painfully apparent. I'll admit there are days when the growing anxiety about the state of the planet and our capacity to feed people as soils are degrading and the climate is changing can feel overwhelming. But the joys of this work have deepened as well, in ways that help me face these sobering realities. I've found that working with plants provides me with a unique sort of inner peace. Being part of a global effort to breed these promising crops gives me purpose and community. And while this all might have seemed like a long shot when I started my work, modern breeding techniques provide reasonably likely paths to develop perennial grains in time to avoid the worst of what might be in store for people and the planet.

I See More Life than Ever Before

Wendy Johnson

When I walk through a field of Kernza, it feels alive, and all of its own volition. It doesn't need the pampering demanded by a summer annual crop like corn, and it doesn't need the crazy amount of inputs required to grow commodity crops either. It is so robust and resilient that I can graze it in the spring and fall and harvest the grain in the summer, all in the same year. In a downpour, my Kernza soaks up excess rain, even the water running off my neighbor's fields. And when there is no rain in sight, it keeps on growing because of its deep fibrous roots. I can watch the birds nest in it and raise their young all summer long. And I can grow a sweet and nutty flavored grain that can be cooked whole, milled into flour, or rolled.

Kernza is just one of many things my partner and I raise at Jóia Food & Fiber Farm, tucked into a glaciated county of northeast Iowa. We also rotationally graze sheep and cattle, and we sell humanely raised pork, chicken, and turkey to local eaters. Recently, we've added a wool fiber business and a 1-acre apple, pear, and chestnut orchard. It's all part of our goal to gradually make the farm more perennial.

I sometimes find myself apologizing or making disclaimers to visitors who come expecting a farm with a white picket fence and turf-like pastures. To most folks, our farm just looks like a jumble of grass and trees until you look and listen closer. In the summer, you'll see some plants more mature than others, some just growing, some just beginning to tiller and leaf. To the untrained eye—or the eye that has been trained by an increasingly standardized rural landscape in this part of the world—it looks like we might have made a mistake. But when I look out at the farm, I see more life than ever before. I see nutrient-rich food for our animals and hearty grains in the making. I see habitat for grassland birds and other winged aviators. I hear the trilling of frogs and toads and the busy movements of furry ground creatures. Each day, I catch little glimpses of the ancient synergy among soil and perennials and ruminants. And I realize we're headed in the right direction.

Back in the early 1970s, my parents were just starting their farming career on this land. They raised hogs in A-frame huts, my grandparents raised cattle, and together they grew corn and soybeans. These were the years when Nixon's secretary of agriculture, Earl Butz, famously told farmers to "get big or get out," words that were backed up with policy. With the help of my grandparents, my parents grew the farm by building the first confinement hog buildings in the county.

My grandparents finally paid off the farm where they lived and retired in 1979, after forty years of payments. And just in the nick of time: The 1980s Farm Crisis followed shortly thereafter, bankrupting many of their neighbors. With a diversified farm business that was debt free, my parents were able to squeak through.

Every time I tell my grandparents' story, I get emotional. The fact that they worked their entire farming career to pay for one parcel of land makes this place incredibly special. They farmed with the focus on the generations to come. I think their story speaks to something I would like to replicate. But it wasn't always that way.

While I was growing up during those difficult years in the eighties, farming didn't look like a viable career path for me. Instead, I moved to Los Angeles and pursued a career in the fashion industry. I even spent a couple years in Brazil. All told, I was away for eighteen years. But when my grandmother died, I reconsidered what this place meant to me. For so many of the kids I grew up with, farming wasn't an option anymore. Their family farms were long gone, lost to the wave of consolidation and industrialization that had washed over Iowa like a typhoon. But I still had the opportunity to come home and steward the land my grandparents had bought all those years ago. And I felt a responsibility to do it well.

What really brought the importance of farming home to me, as I attempted to sell my surfer boyfriend (now partner) on a move to Iowa, was the deepening climate crisis. During my years in Los Angeles, it was an all-too-common sight to see the mountains on fire, with ash falling from the sky and blanketing the sidewalk. I dutifully followed water use restrictions but felt helpless. When I lived in Brazil, I witnessed thousands of hectares of rainforest being destroyed for cattle and soybean production, and I began to feel hopeless. When I moved back to Iowa in 2010, climate change was right in my face again, in the form of massive flooding events that washed away millennia-old topsoil in a matter of hours. But this time I knew what I could do about it. And I felt optimism.

In 2013, I started to transition 25 acres to organic. As I got the hang of this management approach, I continued to transition 20 or so acres at a time, learning as I went. Within the first five years, three things convinced me that annual crops alone were not going to cut it on our farm. Two of those lessons came the easy way. While taking organic cropping systems classes at Iowa State University and attending field days hosted by the Practical Farmers of Iowa, I learned that planting perennials like alfalfa for a few years can provide critical biological weed control in an organic rotation and add "green manure" nitrogen for a subsequent corn crop. At the same time, I realized that if I was going to get our sheep out of a feedlot and back on the land, they would need something to eat.

The third lesson came the hard way. In 2016, we had over 8 inches of rainfall within a few hours. Tile, creeks, and rivers were all overflowing, and our tilled organic fields with small corn and soybean plants were vulnerable. I saw soil wash away and plants covered in a lake of water. The same thing happened in 2018, and I swore that I would not let it happen again.

You could say it was a bit of a perfect storm that created an "aha" moment for me. Perennials help keep the soil in place, even through flooding and drought. I had just been handed an opportunity to sell grass-fed sheep and lamb to a few buyers, and my sheep needed perennial forages to eat. Here was a farm-based business that wouldn't require any fossil fuels, tractors, or tools. It was a win–win all around. And when Kernza came around as a cultivated perennial crop that could also be grazed, I knew that it would be a good fit for us.

But while the ecological case for perennials seemed simple, the economic picture was a lot more complicated. Broadly speaking, restoring land to perennials takes capital, and being an early adopter has meant that I've often been one step ahead of emerging support programs. For many years, I financed the shift toward perennials and diversity on my farm with an operating loan, income I earned from our family's conventional farm, and an off-farm job. Getting Kernza to market also requires infrastructure: seed cleaning tools, dehulling and bagging facilities, and value-added processing. And of course, you need a critical mass of people who know what this crop is and want to buy it, or else it just sits on the farm in a grain bin. There's a lot to do and learn in order to sustain my livelihood as a producer.

One of my responses to these challenges has been to take an active role in the Perennial Promise Growers Cooperative, banding together with dozens of other Midwestern farmers to take our Kernza to market as a group. Not that this is easy either: Learning about grain handling licensing, creating contracts, finding buyers, marketing a grain no one has heard about—and all on volunteer time—is hard to fit into a farm schedule. But it's inspiring to work with other farmers and organizations toward the same goal, and I don't feel like I'm out here all by myself growing a crop no one has heard of. One of the co-op's members and farmer advisors from Iowa heard about Kernza from his daughter and was so intrigued and impressed by the grain that he started growing it—and then invested in a small cleaning and dehulling facility on his farm to clean not only his grain but others' too. It is this kind of passion that helps propel Kernza forward.

But we can't change the agricultural marketplace with the deck stacked against us. So I've also gotten increasingly involved in policy conversations, including as federal agriculture policy co-lead with Climate Land Leaders. I have invited senators and representatives to our farm to share how I have created not just one but two, soon to be three farm businesses using conservation practices. I have given tours to a few staff members, hoping they will share what they learned with their bosses. I tell them that farms like mine have a multiplier effect: I am actively participating within my rural economy, using our local welding shop, farm store, feed mill, veterinarians, and other local services. And I am producing food and fiber for my community, helping it become more resilient in the face of supply chain stressors like what we saw with COVID-19. I tell them to remember this next time they're voting on the Farm Bill.

The Farm Bill is the colossal piece of legislation that provides the roadmap for agricultural policy in the United States. The subsidies in this bill shape which kinds of farms can be profitable, and indeed, which kinds of farms will survive. For decades, the Farm Bill has focused narrowly on production and efficiency, at a huge cost to rural communities and the environment. I would like to see increased funding in the conservation and rural development titles in the next Farm Bill, to create a level playing field for all farmers of any scale and size. I want to see policy that promotes diversification beyond annual row crops, into perennial grains, fruit and nut trees, and vegetables, with animals integrated into farms rather than warehoused in feedlots. I want to see more new and emerging farmers too, and that means we'll need incentives and support for equitable land transition and access.

Sometimes I ask myself whether a farmer like me is really the best person to be working on federal policy and helping develop a fledgling business. But it's my love for this farm and place that inspires me to do this work. I call this place home not only because I was born and raised here but because I feel called to listen to this land. And it needs our help. I remember walking out into the fields one summer when we'd had no rain at all in April and May. The corn had withered in the heat of summer and turned yellow prematurely. I dug 4 feet down into the ground, and there wasn't a hint of moisture in that soil. Then the next season we received more rainfall in the month of May than we did in the entire previous year, and we witnessed extensive flooding, erosion, and smothering of young plants.

The escalating extremes of these weather events are blatant signs that we as farmers need a better plan than depending on increasingly expensive crop insurance and federal commodity safety net programs. It is like continuing to rebuild a house in the same place after hurricanes destroy it year after year. We need a long-term strategy that will not only keep us in the business of growing food and fiber but also be a major climate solution. I think diversified perennials are that strategy.

Perennials give me so much hope, because I know they will continue on. If we plant them, it feels like pieces of ourselves are within them, as if we will continue on too. And I think innately, those perennial seeds are all within us, an ever-cycling seed bank of todays and tomorrows. We can't return to the climate of the past, and there's no guarantee we'll succeed at stopping the worst impacts of the present climate crisis. But at least we'll know that we believed in something that will return year after year after year, that will provide food and help clean water for many. And that is all we may need to feel at peace.

Going to Market

Colin Cureton

I first met Kernza perennial grain at a deli counter. One average weekend morning in about 2010, I woke up and was hungry, so I walked across the street from my partner's home in South Minneapolis to a cafe. Their menu touted a new local ingredient that caught my eye.

While I waited for my coffee, I picked up a bookmark from the stack by the register and learned that the grain in my pastry had amazingly deep roots and, unlike other grains, came back after harvest. This turned out to have a lot of benefits for soil, water, climate, and people. Truth be told, despite being only one generation removed from a central Illinois family farm, I had not thought much about what it meant to convert hundreds of millions of acres of tallgrass prairie and oak savanna to monocrop corn, soybean, and wheat fields, tilled up year after year. Somehow this pastry I was consuming was teaching me about all this. Very interesting, I thought. I left breakfast assuming Kernza now existed for people to eat, easy as that. Little did I know!

Around that time I was beginning what would turn out to be my life's work in food system change. A severe bout with a chronic autoimmune disease of the digestive system had almost killed me as a teenager, leading me to question the wisdom of all the chemicals and processing involved in the food supply chain. As an undergraduate, I traveled and learned about food systems around the world and was introduced to the many complexities and conundrums of trade and economic development. Then I found myself back in the Midwestern United States, wondering what to do about all that.

I spent the next ten years in diverse roles spanning the food justice movement, from policy to philanthropy to community organizing and planning, and somewhere in there earning two graduate degrees. To this point, I had worked mostly at the local level: on individual gardens and farms or coordinating programs across neighborhoods, cities, and regional food systems. But as these dots started connecting, I came to think about this work on the scale of ecosystems and cultures. One of the most profound ways societies interact with the environment is through agriculture. Across the Midwestern United States, we were growing only a few annual crops, mostly to feed cars and animals, not people. We were leaving tens of millions of acres of formerly diverse prairie bare and lifeless over half the year. Why? How did we get here? Could it be different? I had also seen the many ways in which the food system can be reshaped by people working together. I wanted to drive forward agricultural change for the better on the largest scale possible, in the place I called home. One such opportunity happened to emerge.

I came back to Minnesota in 2019 as the newly concocted Supply Chain Development Specialist for the University of Minnesota Forever Green Initiative. Forever Green had been hard at work developing and commercializing new perennial crops—including Kernza—to address all those challenges the bookmark in the cafe told me about almost ten years prior. As I quickly learned, those earliest of early bakery items at my neighborhood cafe had been ahead of the curve. Since then, several trial products had been released by local makers and large food companies. And visionary scientists had developed a new variety of Kernza: a perennial grain domesticated from intermediate wheatgrass, a Eurasian plant brought to the United States decades ago as a forage crop. But there were still lots of questions about how to successfully farm, process, and incorporate Kernza into food products. Researchers didn't have many answers yet for how to successfully move Kernza out into the world, and just about everyone involved experienced challenges. My job, alongside the two other people hired concurrently, was to tackle those challenges head on. How could we leverage the capacity of a land grant university to move this plant out of the research plots and into farmers' fields and grocery shelves?

I'm neither a scientist nor a farmer. I had never taken a class in plant breeding, agronomy, food science, or agricultural engineering—nor had I seeded or harvested an acre of grain myself. Yet there I was, mediating between a university R&D platform and farmers, industry, government, and many communities who were meeting Kernza. I needed to understand genomic selection, plant breeding, yield, test weights, clean-out rates, screen sizes, fanning mills, shaker tables, dehullers, germination, purity, tetrazolium tests, noxious weeds, certified seed, planting rates, combine settings, and converting kilograms per hectare to pounds per acre to bushels to truckloads. I had a lot to learn about agriculture—still do, and always will.

Just four months after I was hired, I handed off the University of Minnesota's first variety of Kernza seed, MN-Clearwater, to ten carefully selected Minnesota growers who had been following the crop's development. We took great pains to communicate the ground floor stage of the crop and market. We did not yet have a comprehensive strategy, and there was very little commercial supply or farmer experience with the crop. But that summer, peering into the back of a grain truck full of Kernza on a late summer day in western Minnesota, I saw just how real perennial grains could be.

We continued to hear from the market that supply was still too limited for industry to consider Kernza "real," so again we cautiously scaled up production. In 2020, I sold 15,000 pounds of Kernza seed to about twenty growers through a unique partnership between Forever Green, the local crop improvement association, and a native seed company that had cleaned the crop. Two years of conversations with grower-leaders culminated in the incorporation of a grower-owned and -led cooperative that would come to shepherd roughly a third of all Kernza production in the United States to date.

That fall, I began to learn about intellectual property licensing in agriculture. I was surprised how often choices fell to the moral compass and motivations of just a few people. For a variety of reasons—from risk management, to downstream product quality, to our interests in developing local and regional value chains—we decided to license MN-Clearwater to three regional seed companies, holding a fourth slot open for a grower-owned cooperative or company. Five years later, two of those initial three licensees remain in the Kernza game and account for half or more of the world's commercial Kernza seed supply. I came to understand intellectual property choices not just in terms of who would get the *rights* to certain technologies but also as a key mechanism for negotiating clear *responsibilities* of market actors adopting these crops. Licensing to a small group of seed companies ensured a stable seed supply and didn't put all our eggs in one basket. It offered a little protection to these early adopters taking a big risk while fostering a little competition for them to do it well.

Also in 2020, a new wave of Kernza seed cleaners, processors, and dedicated brands came online. A supply chain started developing. Young entrepreneurs started getting extremely creative with this perennial grain. They developed custom grain cleaning lines, built equipment for Kernza processing by hand, and started making delicious flours, pancake mixes, pastas, beers, whiskeys, and crackers. In just a year of working at Forever Green, I'd come to deeply respect the genius of the scientists who were domesticating a wild grass into a new grain and the farmers who figured out how to plant, manage, and harvest it. And now I came to appreciate a whole new type of genius: the entrepreneur who knew just how to develop a product that tasted great, looked great, and told a story that connected with people and met a consumer need.

Despite all this progress and enthusiasm, not everyone agreed on the direction of this hopeful new perennial grain. Early-adopter growers saw the need to protect Kernza from a corrosive race to the bottom, arguing for controlled production and a premium on the grain that others found untenable. Early entrepreneurs tried to strike a balance between fair pricing for growers, which they deeply cared about, and viable pricing for even high-end consumers. The State of Minnesota saw Kernza as a tool for water quality. Industry mostly saw a promising but risky new crop that might one day help them meet their bold public sustainability commitments. Food scientists were figuring out how to mill, bake, brew, puff, and otherwise transform it. Makers leaned into taste and finding product–market fit, while researchers expounded on soil health. Still others saw in Kernza a new social vision for agriculture.

Yet by 2022, the market was at a relative standstill. Supply was backing up. A global pandemic had been raging, supply chains were disintegrating, and food companies had begun scaling back to core operations and covering the increased costs of existing ingredients. No one wanted to pour resources into new, high-cost, high-risk, sustainability-oriented opportunities. Also, after much debate encouraging otherwise, Kernza had overshot the market appetite and willingness to pay for the product, and demand outside niche uses and high-end consumers largely stood still. Truthfully, I started to feel slightly disenchanted.

Despite these challenges, early-adopter growers, entrepreneurs, industry partners, communities, and researchers did not quit. They kept at it. New energy came into the fold. By 2024, their collective efforts started turning a corner. Several nationally distributed Kernza products were released. Kernza products won awards at major food industry trade shows. Some market channels sold out of grain and flour. Growers saw the clear market signal and planted new acres. This toehold in the market was hard won, breathing new life into the small but growing Kernza community.

During a volatile half-decade marked by drought, floods, extreme heat, pests, pandemics, and political upheaval, Kernza has moved into the market. I have watched the early days of a perennial grain economy take shape through the collaborative efforts of thousands of people— working behind markets and politics, toward promising solutions.

Five years ago there was no reliable Kernza seed or grain supply, no products you could conveniently buy. Now as I write, there are five US distributors of Kernza grain and flour. More than fifty products across the country have incorporated this novel crop, and there are at least three places in my neighborhood I can buy Kernza pasta, pancake mixes, or beer.

Assuredly, not everyone who has worked with Kernza has found success. There have been many frustrations with this grain's early launch onto the landscape and into the market. Let's not sugarcoat it: Being an early adopter is hard, risky, and at times frustrating. These folks take big risks, bear a great load, and often have the scars to prove it. We owe them a debt of gratitude.

I think often of the farmers on their planters on an early September morning, or the entrepreneurs running a load of grain through the dehulling line for the third time. Although Kernza and perennial grains are very hopeful, and there is now a robust perennial grains community, the reality is that it is still a new and challenging crop. At some point or another, everyone working with Kernza is somewhat on their own—on a combine, in a shed, inside a warehouse, at a lab, or on the road—by themselves, just trying to make it work.

Inspired by these risk takers, I decided in 2023 to try my hand at selling some perennial crops myself. I found a business partner, a twenty-year global food industry veteran, and we launched a venture studio focused on taking new regenerative crops to market. So far we have two active ventures: One is taking Kernza to market in Europe, and the other is developing a new hazelnut industry in the Upper Midwest. As I write this, I'm doing the startup hustle, burning the candle at both ends with a full-time job I love while also building the early days of companies that can act more directly and decisively.

I have a running joke that plant scientists and others in this work tend to reflect the characteristics of the crops they work on. Maybe people with certain traits are drawn to crops well suited to their person- alities, or perhaps those plants and their agroecosystems make us more like them. For example, the hazelnut people are somewhat quirky, take years to get to know, and are unbelievably patient. Meanwhile, the folks developing cover crops tend to be more conscientious pragmatists, more engineers than artists.

So, who are the perennial grain people? They are steadfast prairie people, resilient and deep-rooted. Like Kernza, they can take a storm. They see a far horizon. They are fundamentally good people, and once planted they tend not to go anywhere for a while. They create habitat for others, in ways seen and unseen. They are good hosts. They make wherever they go better.

Over the last five years, I have become a bit more like the prairie myself. A little more quiet and humble, with a deep and dense network under my feet holding me up. I've learned that I can survive a political or economic drought and can absorb the work equivalent of a few-inch rain. I'm thankful to be doing this work, to be building healthier soils, proverbially and literally, for my children and generations to come.

Now, about fifteen years after I first met Kernza, I hand out the very same bookmark that first introduced me to Kernza—at field days, community events, and the State Fair. I watch people have the same realization of deep, dense roots, what they can do for soil and all the life that depends on it. And our collective effort to bring a perennial vision to life continues.

Field Notes from a Perennial Kitchen

Beth Dooley

As I write this from the porch of our family cottage on Madeline Island in the western end of Lake Superior, a hazelnut pound cake cools on the counter, elderberry–raspberry jam simmers on the stove, and that yeasty-toasty scent of freshly baked focaccia fills the kitchen. For most of my adult life, I've been the house cook. But a bag of locally milled Kernza flour inspired my daughter-in-law Alli to make the dense, buttery pound cake with my grandmother's recipe. Early this morning, our middle son, Kip, picked elderberries and what was left of our raspberry bushes, while our youngest son, Tim, kneaded Kernza flour into his dough. Matt, our eldest, promptly shooed me out of the kitchen when I started to whisk a hazelnut oil vinaigrette. "Mom, we know how to do this," he said, a refrain all mothers must eventually, if grudgingly, learn to accept.

As a cook, food writer, cookbook author, and mom, I've focused my recipes and articles on the flavor and nutrients of our ingredients. Over the years, meeting with farmers, researchers, food scientists, and producers, I've come to the humbling conclusion that good food relies on good soil. Complex with diverse microbial activity, fertile soil is alive as you and me.

The act of cooking has been a key nutrient in the soil of my own life. As lucky as I am to write food articles and recipes, it's when someone asks "What's for dinner?" that I truly come alive. I can then close my laptop and head to the stove and merrily chop and sizzle and simmer. Food is no longer words on the screen but an aromatic and sensual experience. Understanding where the ingredients come from enlivens the meals I serve.

The Kernza flour for Tim's sourdough bread came from a field day on Carmen Fernholz's farm in Madison, Minnesota. There on a sweltering August afternoon, I joined a group of farmers, food processors, bakers, and chefs and looked into a soil pit dug into a shimmering green field where Kernza's thick roots reached 8 feet into the ground. Carmen relayed the benefits of this crop that, like backyard grass, returns each year and does not need to be tilled up or replanted like annual wheat, a major cause of soil erosion and nutrient runoff. Kernza's deep roots capture carbon. Meanwhile, its seeds can be milled into flour, brewed into beer, and distilled into liquor. Kernza is a regenerative crop, Carmen explained. Unlike corn and soy, monocrops, it has as much value to the land as it does as food. "Growing good food and providing farmers with a reliable income should not be mutually exclusive," he declared.

Kernza flour is different from wheat. It has flavor, an attribute we do not associate with white, all-purpose flour. It's more distinct than whole wheat, and its texture is slightly different. With its grassy, nutty notes, the flavor reminds me of rye but is less bitter. Its texture feels slightly gritty. Kernza bread dough tends to be looser and has less structure than wheat dough. The loaves are dense and moist. When making bread with Kernza, it's best to mix in at least 45 percent wheat flour, or more for a lighter loaf. Tim, our bread-baking son, uses Kernza to feed his sourdough—it seems to activate it immediately.

Substituting 100 percent Kernza for all-purpose flour in my grand-mother's pound cake and shortbread recipes was a delightful surprise. Kernza absorbs butter differently, producing a richer, moister, distinctly buttery cake. With brown sugar or maple sugar, the shortbread develops a caramel taste. I've also made loaves of Irish soda bread with my mother-in-law's recipe, using 60 percent Kernza for a texture that reminds me of the dark rough loaves we ate with sharp Irish cheddar cheese and a good dark stout.

The hazelnuts for Alli's pound cake came from the American Hazelnut Company, a farmers' cooperative that processes toasted and seasoned hazelnuts, presses the nuts for a golden culinary oil, and then grinds what's left of the pressing into a gluten-free flour. Native to both the Midwestern United States and Ireland, hazelnuts capture water, shelter pollinators, and provide forage for animals.

When I worked on *The Sioux Chef's Indigenous Kitchen* cookbook with co-author Chef Sean Sherman, three-time James Beard Award recipient, I learned that these tall, shrubby trees produce nuts that are tinier than their Irish cousins but with a richer, more distinct flavor. Hazelnuts have long provided Indigenous people food and fiber, wood for baskets and containers, forked twigs for divining rods to find water, and more. In my own kitchen, I've found that they make a flavorful addition to baked foods (especially those with Kernza flour) and are delicious tossed into salads and eaten as healthy snacks.

The berries we harvested ourselves. When we're on Madeline Island, in August, nearly every meal includes berries—blueberry pancakes, blackberry muffins, raspberry–elderberry jam. Elderberries are the bluest fruit of all, sporting lacy white flowers early in the season, followed by clusters of tiny, dark fruit. Native elderberries grow with abandon in the wet ravines near our cottage. The berries are extremely tannic, tasting like an unripe apple or harsh red wine, a flavor that can be tempered with honey or maple syrup, especially when combined with raspberries into syrup and jam. The benefit of the bushy elderberry trees to wildlife is immeasurable. The pollen-heavy flowers attract a wide variety of bees, flies, and beetles, and the leaves provide food to moth and butterfly larvae. Birds love them for perching and nesting. Planted as borders on farms, they stem runoff, capture carbon, and nourish the soil.

This hearty perennial is having a renaissance thanks to its natural healing properties: Find elderberry cough syrups, vitamin pills, health drinks, and kombuchas on supermarket, co-op, and drugstore shelves. A few years ago, I noticed the label on a bottle of elderberry tonic— Midwest Elderberry Cooperative—and then reached out to one of its growers, Natasha Simpson, of Regeneration Acres in Clayton, Wisconsin. "I like growing all these vegetables for our CSA—the tomatoes, lettuces, carrots, and potatoes," she told me over the phone. "But there is something really appealing about growing perennial crops. You have a relationship with the plant that's permanent, not so transient. I've always hated that at the end of the season, I have to dig up all the healthy vegetables and herbs I've been nurturing because they'll die at frost. But elderberries are native, I don't have to baby or pamper them, they're resilient, and they like it here." Later tonight, for dessert, we'll serve the pound cake topped with that elderberry–raspberry jam, a bit of whipped cream, and a scattering of toasted hazelnuts.

Now that I'm closing in on the final chapter of my working life, I'm thinking of what I want to leave behind. I'm remembering the lessons of those researchers, farmers, and chefs working to regenerate our food system and, in doing so, regenerate our relationship to the land and to each other. Just as they are learning from all the complex activity in the soil, we too can learn about these new and rediscovered foods—baking bread, simmering soups, putting up jams, and whisking vinaigrettes.

As I grapple with the challenges we face and try to understand how to become a better steward of our natural bounty, I cannot help but rejoice each time I see my sons and their partners washing and chopping what they have pulled from the soil, what they've kneaded and baked into bread. Come into the kitchen. Let's cook!

Hand Threshing

Aubrey Streit Krug

The sunflowers reseeded themselves into a forest to sway in the wind.
I had to look up to see the bees.
After petals fell each head bent with weight into a shepherd's crook.
What I harvested was a secret even to me.
The dry heads held their yield tight against thump and crumble.
Facts like songs only sometimes come free easy.
The wind teased help and skittered light chaff from grackle black seeds.
At last I could see every necessary sort of thing.

Generation by Generation

David Van Tassel

The plant in front of me has two stout green stalks shooting up. At the top, level with my eyes, the stalks culminate in clusters of drying seed heads—and a few late blooms resembling miniature sunflowers. My eyes are drawn to the top half of these two stalks, where a handful of beautiful dark green leaves hold fast, stiffly perpendicular to the stalk or angled upwards. But these leaves' beauty is marred by dark bumps—signs of "rust" fungi mycelia—and brown polygons, signs of "blotch" fungi similarly feasting on the leaf tissue. Other stalks on this plant have not fared so well. Three have just a few blotchy dull green leaves, while the other five or six are withered and brown, with only a few crunchy brown leaves hanging on. I scribble some numbers in my notebook: scores for stalk health, leaf health, overall plant shape. Then I move on to score the next wretched, defeated plant.

I've scored about 700 plants in the last forty-eight hours, and I'm excited to see that we have strong, uniform disease pressure in each plot. This gives me more confidence that we're getting sound data and the survivors actually have some genetic advantage. Last year, the diseases

seemed patchy, and our data didn't show solid evidence of genetic variation for resistance. Probably some of the healthy plants were just lucky;
spores didn't happen to land on them, or at least not until much later.

Although it's a good year for resistance breeding, I feel sad watching
wave after wave of disease hit the plots. It's not that it's unusual for the
stalks on this plant, *Silphium integrifolium,* to die back. Its strategy is
to keep roots going year round, getting woodier and thicker over the
years, while the stalks are more of a seasonal phenomenon. This combination of deep woody roots plus stalks that can be mowed or burned to
the ground each winter could give farmers some of the drought-toler-
ating benefits of deeply rooted tree crops, without the need for ladders
at harvest time.

But we expect the stalks to die back after a hard freeze, not at the
height of the summer, when sugars produced by photosynthesis in
the leaves are needed to recharge the roots and grow seeds that can be
pressed into calorie-dense oil. It's hard not to feel some sympathy for
the disease of these plants as fungi spread from leaves to stalks—plants
that I'd been so proud of, that showed such vigor and promise early in
the season, plants that came from seeds I'd hand pollinated, harvested,
threshed, weighed, labeled, germinated.

Silphium is a native prairie species, but I didn't grow up on the
prairie. I grew up in a big, crowded city, fascinated by the subtropical
plants around me. In high school I imagined working in international
agricultural development. Yet the more I learned the less confident I
felt that I, as a city kid, could really help farmers anywhere. I chose to
study botany and hoped to be able to use that discipline to someday
give farmers everywhere new options (instead of advice). I didn't have a
clear idea of what those options should be.

Then, my wife's Kansas relatives introduced me to The Land
Institute. Compared with the plant biology PhD program I was
finishing up at the University of California, Davis, where I was

contributing to incremental change in botanical knowledge, the idea of creating whole new perennial crops seemed so much more exciting. In my early days working on perennial oilseeds, I imagined finding that one-in-ten-thousand rare mutant with a seed head twice the size of any other.

That was before the pests and diseases struck in full. Now I see a tangled web. My job is to put my thumb on the scale in favor of the plants in their fractious interactions with pathogens, generation by generation.

It's coevolutionary work. We are forming a new relationship between humans and silphium. Through accelerated evolution—plant breeding—genes that were once very rare in the wild will become common, leading to changes in seed size, shape, composition, and germination. Overall, humans will take over the job of moving seeds to new environments (so silphium can lose the "wings" that help seeds to drift away from their mother), making harvesting easier. We will nudge the plants toward taking more risks: germinating seeds quickly and uniformly, investing in sexual reproduction even when conditions aren't perfect. In exchange, we will hold seeds in safe places to be able to replant in case of disaster. And we will help find rare disease resistance genes from different places and bring them together to keep one step ahead of the fungi.

Through this new relationship, we will gain a new food: oil-rich silphium seeds that taste like sunflower kernels.

Our research is well under way, in my lab and with a range of collaborators. We've learned how to pollinate and germinate silphium. We have experimented with strange tools and machines for harvesting and cleaning silphium seeds, which are not handled well by existing farm equipment. However, to further this coevolution, we have a lot more work to do, including honing the techniques we use as plant breeders.

Recently, I visited colleagues at the Donald Danforth Plant Science Center in Missouri, and we scored disease in the field. We also snipped off the leaf tips of hundreds of seedlings in the greenhouse. Each leaf piece went into a small manila envelope and from there into a freeze-drier. Those crunchy dried leaves will someday be sent to our collaborators at the Hudson Alpha Institute for Biotechnology in Alabama, where DNA will be extracted. Eventually we will be able to build a statistical model connecting DNA variation with plant health. We should then be able to DNA fingerprint a new batch of seedlings and immediately predict which ones are the most vigorous and disease resistant. The prediction won't be anywhere near perfect, but the model can be updated as we later expose those seedlings to actual disease and multiple environments. My colleague Lee DeHaan has shown that genomic selection accelerated genetic gain in Kernza, and we have high hopes that this version of artificial selection can help us ratchet up silphium's ability to slow and contain the damage from several diseases.

This new approach could speed up our breeding, but successfully planting and harvesting silphium at the farm scale will not be easy. We will need to invent methods and equipment to do this, and farmers will need to learn and adapt. For even a few farmers to start growing silphium, no matter how imperfectly, we will need resources for solving problems with machines and markets. It's hard to get funding for something that isn't a crop yet.

That's where culture matters. For people to value silphium's potential as a future food, they'll need to meet silphium in the first place. We have begun using civic science projects to introduce silphium to people, to test silphium in hundreds of backyard environments, and to help people learn about domestication and sustainable agriculture. Eventually, if silphium is to displace annual oilseeds to provide humans with needed calories and nutrients, we will need to create new recipes and change our supply chains.

Indigenous communities on this continent have long known the *Silphium* genus, tending and using these plants as part of their cultural practices of stewarding prairies, although we haven't learned of any human groups eating silphium's large seeds before. Initiating a domestication relationship with silphium means we have the chance to consider the wide range of reasons people might care about this future oilseed crop. Silphium could provide nectar for bees and nutritious forage for livestock along with high-quality oil for people. New and renewed webs of relationship with silphium, in North America and beyond, could help sustain the domestication process for the long term.

I can imagine the joys and sorrows of silphium entrepreneurs and eaters in the coming years. I suspect there will be scenes of cursing while struggling to un-jam brittle silphium stalks packed in impossible-to-reach parts of expensive harvesting equipment. I picture someone running silphium's weirdly shaped oilseeds through a series of blowing and sieving machines again and again after the kids are in bed, striving to get the grain clean enough for the buyer the next morning. I see people tinkering with an oil press, waiting anxiously for the seeds to release their precious yield, then tasting and comparing the different batches. These heroic early adopters will need to sustain themselves with their own daydreams of hungry cattle shoving each other to be first to that bale of perfectly cured silphium or a check from the company experimenting with novel ingredients—another early adopter bravely betting that their customers will pay for a story of pollinators and deep roots cleaning the groundwater. Lots of people will need to take risks, and I hope there's a way that the rest of us can support them, just a little.

After the Flood

Piyush Labhsetwar

In early July of 2023, I woke up to the steady rhythm of heavy rain. Still half asleep, I got online to check the water level of the river. Sure enough, it was shooting straight up. We'd already received more than 4 inches of rain, and the river had more to rise. I immediately drove over to the community farm to check the research plot I'd planted the previous summer. I had to take a different route than usual, since some of the roads were already closed, but I needed to learn what had happened to my pawpaws and perennial grains.

What I saw there made me gasp. The river had jumped its banks and was flowing with full force. The water was over 5 feet high, and it had brought piles of wood, leaves, and trash with it. I waded in to help another farmer rescue some of their materials and wondered about my own plants, hidden under the water. Would they survive here?

I had chosen this plot next to the river precisely to conduct an experiment in flood-resilient agriculture, and I'd selected the species and layout accordingly. As a perennial grain researcher, I decided to plant Kernza along with a potentially perennial wheat, resulting from a cross between intermediate wheatgrass and durum wheat. For a companion tree, I chose pawpaw, given my love for its tropical taste. These small trees, which can be found in the understory of many forested parts of the eastern United States, produce the largest edible fruit native to North America. I planted 100 of them in the shape of a cross, with rows of perennial grains at each corner, so I could study their interactions. Together, I hoped the trees and grasses could help hold the soil under the pressure of flowing water.

When I returned the next day, I saw Kernza and perennial wheat plants bent over, leaning in the direction of the flow of water. Stressed, but not destroyed. The pawpaws, amazingly, were unfazed. The tree tubes protecting them mostly held up. Save for a few gashes here and there from the debris, they looked healthy, and in the coming days and weeks I would in fact see them growing as if nothing had happened. Meanwhile, I spent my time weeding the plots, joined by groups of volunteers. We kept a close eye on the perennial grasses.

But when I took a break from weeding and stood up to look around me, the most interesting things I noticed were the new deposits of silt that had arrived with the floodwaters. My partner and I walked the land and noticed that these wavy new patterns of sediment were drying at different rates, signaling differing compositions of sand and silt. Soil tests revealed that these new deposits were packed with organic matter and micronutrients. Grain harvests were minimal that flood year—those plants had been through it—but after mowing, the perennial grains came roaring back in a lush green reflection of all that our soil had gained.

I was reminded that floods had long been a part of life on this land. They used to happen quite predictably and were called spring freshets. Flood deposits helped build deep loam soils, making the land a fertile honey pot. I thought of the Indigenous Nonotuck, Nipmuc, and Pocumtuk peoples, who have known this dynamic floodplain as Cappawonganick. More recently it also gained the name Mill River.

Only in very recent history have the humans living here and in other floodplains become disconnected with seasonal shifts in riverbeds, funneling waterways into predictable channels. We fear floods and avoid them. With septic tanks, centralized wastewater treatment plants, and historical pollution from industries, floodwater is presumed to be contaminated, and farmers have to discard any produce it touches. But floods do not just bring devastation—they also bring renewal in the form of new soil matter. The trouble is that with climate change, floods are happening more frequently and unpredictably. In less than a year after this big July flood, the community farm land flooded three more times! So we have to learn to live with floods in a new way.

Like the new silt, I settled in this valley a handful of years ago. I'm a new migrant to this continent from South Asia, where I grew up disconnected from land. My journey of reconnection started when my partner introduced me to Mother Dawn, steward of Randolph Street Community Garden in Champaign, Illinois. Turning the soil, moving with the seasons, and growing vegetables, I felt new possibilities of allyship with the land. Soon after, I sought out an opportunity to work at The Land Institute in Kansas. I did a deep dive into plant genetics in support of perennial wheat breeding. After almost four years of transitioning my professional work toward agriculture, I felt a strong connection to these grains. So when I moved to Northampton, Massachusetts, following my partner, I carried seeds with me, hoping that both these plants and I would be of some use in our new home.

The perennial grains and pawpaws have indeed rooted together in the soil here, braving weather shocks, like late freeze and drought, and a series of floods. These days the Kernza plants are growing tall, while the pawpaws are ready to peep out of their 5-foot tree tubes. As they mature, they will help to slow down future floodwaters, safeguarding the life around them.

I've come to learn how the human community can be like these perennial grasses, working with each other and with this place so that folks like me in the diaspora can find our footing. Over 150 years ago, this farmland was the location of an abolitionist utopian society and a terminus to the Underground Railroad. In recent years, local food activists protected this land from development and set up a community farm for folks to grow food, flowers, and livelihoods—in the face of all the forces that displace people from the land around the world.

Now, I am entrusted with stewarding this 121-acre community farm called Grow Food Northampton that includes my research plot. I do this alongside hundreds of gardeners and dozens of farmers. We offer affordable farmland and some housing to smallholder farmers, prioritizing those from historically marginalized communities. We will continue to try to repair past harms on and around this floodplain. I'm sure there will be more surprises and struggles. But together with this community, I'm committed to the work ahead.

A few years ago, my parents visited from South Asia. I brought them to the community farm, where they watched me check on the pawpaws and grains in my plot. I could tell they were reconciling the fact that their hoped-to-be-white-collar-worker son was returning to the fields. They had raised me away from our ancestral hometown in a place where people spoke a different language, and our lives were not directly connected to land in the way they would have been in previous generations. So I have felt like an outsider since childhood. But I am starting to grow a sense of belonging to this floodplain. My mom, joining me next to a pawpaw seedling, exclaimed, "Even after you are long gone, people who you haven't met will enjoy the fruit of your labor." These perennials will keep me with this place beyond my mortal time horizon.

Shallow Roots Run Deep

Tim Crews

I look out the window of the Amtrak train I am riding between Kansas and Washington, DC, headed for a meeting on extreme heat episodes caused by climate change. Moonlight illuminates vast, plantless landscapes of bare soil waiting for winter wheat to be sown later in the fall. I know as we cross the country, the vegetation will shift to corn and soybean with the harvest in full swing. I think that to most people, this landscape conjures up feelings of orderliness, like a freshly cut lawn. Rich smells of fecund soil. Comfort. To farmers and farming communities the feeling of orderliness extends beyond aesthetics. It's a brief but uplifting sense of victory over always-encroaching weeds and the satisfaction of completing hard, timely work. A sense of relief that seeding can soon happen; a sense of anxiety about what the weather will bring next. That this landscape is inextricably linked to food I think explains to a large degree why it makes us feel good. Even though nature has been subdued.

Only about 10 percent of the prairie land that covered the North American landscape at the onset of European colonization exists today.

And yet, even though most of the prairie is long gone, we still depend heavily on that historical landscape. Yes, true story! The rich Mollisol soils of the US breadbasket were *formed* under diverse, perennial plants of the past. Over thousands of years, the deep roots of forbs and shrubs, and the especially prodigious fine roots of grasses, all grew and died and decayed and grew more. They fed a huge assortment of microorganisms, cycled nutrients, and built a thick, dark topsoil.

Agriculture is not only indebted to perennial grasslands for the essential natural capital of soils. The plant communities of almost all natural ecosystems build fertile soils in a similar way. I remember an old postcard: "Gravity, it's not just a good idea, it's the law!" It took me close to half my life to realize that perennials were in fact the law of the land. I think the countless times I spent escaping from Albuquerque as a teenager into the hinterlands of New Mexico, Colorado, and Arizona laid the foundation, an unconscious understanding of what natural vegetation looked like. That same passion for hiking has continued, introducing me to plants in coastal rainforests, alpine tundras, deciduous forests, tropical rainforests, cool and warm deserts, short- and mid- and tallgrass prairies, scrublands, and savannas. All diverse, productive, and overwhelmingly perennial.

There is surprisingly little in the scientific literature specifically evaluating the relative extents of perennial and annual vegetation across different natural ecosystems. However, in 2023 three Israeli botanists published a study in the journal *Nature* in which they estimated that across all ecosystems on land, only 6 percent of plant species are annuals, whereas 94 percent are perennials. They also pointed out how dominance of annuals has spiked during the Anthropocene epoch, due to humans replacing natural ecosystems with annual crops.

There is good evidence that early Neolithic farmers started the first agricultural revolution in small areas where natural disturbances like floods or fire cleared vegetation, allowing for seeds of annual crops to be sown. With time, our human ancestors shifted from appropriating disturbed areas to facilitating new ones. The surface area of Earth covered by annuals has grown in proportion to the human population over the last 10,000 years, with a step-change taking place around 1800 when fossil fuel energy turbocharged our capacity for land clearing. In the last 225 years, our population increased from 1 billion to 8.2 billion. Annuals are more common, and more soils are degraded, than at any time since the retreat of the last ice age.

I think the first time perennialism really took root in my frontal lobe was in 1981. I was an undergraduate at the University of California in Santa Cruz when my professor, Steve Gliessman, assigned *New Roots for Agriculture* by Wes Jackson in my agroecology course. That read was a defining moment. In *New Roots*, Wes Jackson, co-founder of The Land Institute in Salina, Kansas, drew attention to the perennial blind spot like few had before, and few have since. And while Australian Bill Mollison and other permaculture luminaries certainly shined some light on the importance of perennials, Jackson took his critique of annual agriculture to a different level, offering perspectives from evolutionary biology, geology, sociology, and even religion. Jackson's critique simmered in my thoughts, rarely at the forefront but always present, as I built a vocation in ecology and agroecology over the next fifteen years.

Around the turn of the millennium, Land Institute agroecologist Jerry Glover asked whether I would like to participate in a research project comparing a handful of central Kansas wheat fields with adjacent perennial prairie meadows being hayed for forage. Both haying and wheat growing had been taking place for at least seventy years. The project, with a multidisciplinary assortment of graduate students from around the United States, would specifically look at yield, energy inputs, soil nutrients, organic matter, and soil food web diversity and abundance.

I will admit, compared with the perennial food plants of the arid Southwest, where I was teaching agroecology students and working alongside Indigenous farming communities, perennial hay meadows seemed less than riveting. Clearly, Jerry was interested in these naturally occurring perennial polyculture meadows as an analog for some future grain-producing ecosystem, but I found the comparison to be pretty ho-hum and feared others would as well. And yet Jerry had a way of getting people like me on board, so I agreed to join.

The field work, sample analyses, writing, and publication of this comparison took maybe five years. And the results were anything but ho-hum. For example, the nitrogen exported in the yields from the hay meadows and wheat fields was quite similar, but the wheat fields could sustain those exports only with annual applications of nitrogen fertilizer, while the hay meadows received none. Moreover, the total nitrogen stored in soil organic matter in the fertilized wheat fields after seventy years of production decreased to significantly lower levels than what was stored in the unfertilized prairie, also harvested for seventy years. Beyond nitrogen, the study underscored how the community of organisms that made a living in the hay meadow soils—such as nematodes and bacteria—were more beneficial to the plants than the community that dominated the wheat monocrops.

Seeing the amazing soil health attributes and ecosystem integrity of the prairie hay meadow—which supplied regular harvests with no inputs of fertilizer—led me to value not only perennial vegetation but the soil ecosphere that it helped create. And I couldn't unsee the juxtaposition with tilled annual wheat fields.

Two years after the study was published, I moved to Kansas to join the staff of The Land Institute. Now, after twelve years leading the institute's soil ecology research and peering belowground into the mighty Mollisols built by perennials, I've seen interest and enthusiasm around perennial grains grow tremendously. Multidecadal projects to develop perennial grain species that can then be integrated into synergistic crop communities are appearing around the world.

In 1995, Chinese professor Tao Dayun and Thai professor Prapa Sripichitt successfully made a cross between an annual rice cultivar and one of its perennial relatives from Africa. The story that followed this single plant hybridization included many different chapters: one starring a high-profile international program, one in which the project was defunded and shelved, and one where an early career researcher dusted off the project and rebooted the breeding program. That last one led to a chapter involving long-term grassroots international support and almost two decades of highly focused research with very modest funding.

In 2018, Dr. Fengyi Hu and his research group at Yunnan University in Kunming, China, officially released the first perennial rice variety to farmers. In 2022, Hu's group garnered global attention by publishing a paper in *Nature Sustainability* in which they described how PR23, a perennial rice cultivar, had yielded eight harvests in a row, two per year for four years, with comparable yields to the annual rice varieties commonly grown in southwest China. Now Hu and his team are expanding their breeding attention to develop upland and frost-tolerant perennial rice as well as a range of flavors and textures like aromatic and sticky varieties. The rice is being evaluated in Southeast Asia and sub-Saharan Africa. Plant breeding is a slow and gradual process, and it is almost antithetical to say, "We have arrived." But the incredible progress made by Chinese colleagues has made history as the first ever high-yielding perennial grain, and it illustrates what is possible with sustained dedication and support.

Buoyed by the success of perennial rice, another effort to breed an important perennial staple grain has recently been initiated in the highlands of Bolivia. Dr. Alejandro Bonefacio is a well-known Aymara plant breeder who for decades has been improving varieties of quinoa adapted to the Bolivian altiplano. Around 2011, a surge in quinoa popularity in the United States and Europe resulted in an unprecedented spike in quinoa demand on the international market. In what Bonefacio characterizes as a hurried and short-sighted response to this demand, many Bolivian farmers took to plowing thousands of hectares of hearty perennial grass and shrub vegetation, to expand the lands available for annual quinoa production. Within a few years, the spike in quinoa demand crashed, in part because of an increase in domestic quinoa production in the north, leaving recently cleared landscapes unplanted and vulnerable to wind erosion.

The ecological impacts of the boom-and-bust quinoa experience got Dr. Bonefacio thinking about perennialism and the consequences of eliminating perennials from agricultural landscapes. Thus, for some time he has been wondering whether a particular viny perennial plant that grows on the arid southern end of the Bolivian altiplano might be crossed with annual quinoa to introduce the perennial habit into widely cultivated quinoa varieties—in the same way a perennial relative of rice was crossed with annual rice back in 1995. A new perennial quinoa project is getting into the ground.

Projects focused on the breeding, ecology, agronomy, climate resilience, and social adoption of perennial grains now exist in over thirty countries around the world. From Kernza and barley in Scandinavia, to silflower oilseed in Argentina, to the legume sainfoin in Türkiye, and perennial sorghum in Uganda, diverse crops and geographies are expanding the likelihood of a perennial transformation in agriculture.

Still, annual agriculture shapes the reality in which we live, and its roots run deep, culturally and psychologically.

Looking out the window of the train, I see the annualization of the land everywhere. Views of bare fields go on for minutes, punctuated by only brief seconds of remnant prairie strips or woodlands. Changing this view will take time. Breeding new perennial cultivars, figuring out how to grow them, where they grow, what plants they like to grow with, who will eat them, how to eat them—it all makes me reflect on how the totality of a transformation of this magnitude will happen only over the course of cultural evolution, not in the timeframe of a few studies or government initiatives.

And what about time? Some colleagues wholly agree that the perennialization of agriculture would be an ecological and social game changer but do not believe there is enough time. They argue that we need faster solutions to address the existential threats of climate change, biodiversity loss, and food security. Yet the technological fixes developed so far fall short of addressing the scale of agriculture's challenges. And no matter what the future holds—a seamless transition to a renewable economy or a profound down-powering and reduction in material wealth—any progress toward an agriculture that reconciles the way people farm with the way nature functions will be an important step in the right direction.

Rooting for Ourselves

Laura van der Pol

what gets richer the more porous
is precious yet
no one wants it in their house
is a glimpse of your future
and mine

Look down.

The back wheels of the truck lift off the ground and the hydraulic motor squeals. Will doesn't say anything as he reverses the lever that directs the probe up or down to extract it from the soil; we've fallen into the habit of not speaking much over the roar from the truck engine and the motor. Words get swallowed in that blast of sound, so we limit them to a few simple phrases: "good," "need more soil," "redo." Will doesn't need to say anything; I know he's hit another rock. He pulls the metal pin that holds the long, steel cylinder about 3 inches in diameter and uses a skinny PVC pipe to push the reddish brown soil onto the metal trough on the side of the probe truck. I admire the contrast of the red ferrous oxide and the white calcrete deposits and rocks that are complicating our sampling efforts. Every core is a surprise.

Although the broad forces of climate are the same for all the soil in an area—the temperature and moisture dictate which types of plants can grow and for how long—each organism, rock, and water drop leaves a mark on a particular place. Just a few hundred feet away, there seemed to be no rocks at all in this field. The soil on the other side of this sloping terrace was a milk chocolate brown dense with clay. Particles so fine and pore sizes so small that the soil clung to itself, sticking to everything—the soil core, the paint scraper we use to cut the core to our chosen depth intervals, our fingers, the sample bag. Yet here, the brown has mostly given way to this red ochre, a coarser texture with plenty of sand. Not to mention the rocks. I wonder whether this area was once a streambed.

In any case, the rocks are annoying. We're hoping for cores that are at least 1 meter long, but so far we haven't managed to find a spot where we can extract soil more than 50 centimeters. We move the truck ever so slightly—a foot forward—and try again. Will brushes aside the vegetation from beneath the probe with his boot, lowers the foot, and drives the probe into the soil once more. This time the probe moves effortlessly down and down; we can't help but smile our relief.

I line up the end of the core at the zero centimeter mark we have drawn in Sharpie on the rusted metal tray and place the paint scraper there. Holding the PVC pipe against the end of the soil core, Will pulls the metal tube toward himself, leaving the soil behind on the tray like it's a giant Play-Doh extruder. I cut precisely at the depths we have marked—60, 30, and 15 centimeters—and then use the scraper and a paintbrush to move the soil for each depth into a labeled plastic bag. Frenetically, I pick out the bits of grass that have found their way onto the tray; we don't want this "litter," as soil scientists often uncharitably refer to plant debris, in the sample. Still, I question whether it's better to move more quickly and leave the debris to be picked out later or expend the extra seconds now. We must finish collecting the samples today, and you never know how many attempts will be needed to get a full sample, whether the truck will break down, or if the weather might shift suddenly.

The last time we were sampling, the probe stopped abruptly with the core tube, full of soil, still partially in the ground. After puzzling unsuccessfully over it for an hour, we decided to send some of our team to the nearest auto parts store for a new battery. Another hour later, fresh battery installed, the probe remained an inert pole in the ground. At that point I started to worry; we could not leave an immobilized truck in this farmer's field, but how could we move it? Luckily Greg, a US Department of Agriculture soil scientist with a couple decades of experience tucked into his boots, was unfazed. After some sighs and fiddling with wires, he found a fuse that had blown, and miraculously, he had a spare in the other truck.

Today has gone smoothly aside from the rocks, but this anxiety pervades field work, whether you're an ecologist or a farmer. Our dependence on favorable weather, equipment prone to failure, amenable soils, and the convergence of many people into one remote place at the same time makes us vulnerable. Every successful sampling event feels like victory, even if it represents just a small drop in an ocean.

What brought us out here today was curiosity: Can we detect differences in soils grown under perennial grains compared with annuals? And how do these soils change over time under each type of cultivation? We have good reasons to think that perennials benefit the soil in many ways, but because the perennial grains we're studying are recent developments, we don't know their particular impact for sure, nor do we know how long changes might take to become detectable.

Soils develop under perennial vegetation over a long period of time—think thousands of years. And yet for us humans to detect these differences in soils under annual versus perennial cover is actually somewhat difficult. For one thing, soils are highly variable and are influenced by many things. It's like we're trying to listen to the way one violinist's tune has changed in an orchestra of musicians—but each musician is playing a different song. For another, soils exist on a different time horizon than people. Less than the width of a human hair (0.016 to 0.05 millimeters) of soil forms in a year. In all the time that's passed since the dawn of agriculture (some 10,000 to 14,000 years), just over a meter of soil has formed, and yet our best estimates are that we've washed more than five times that volume of soil into the ocean after we've exposed it by clearing native perennial vegetation. Chances are you barely think twice about all that missing soil when driving past a field that stretches to the horizon full of a single annual crop, scarcely blemished by herbivore or weed. Your heart may not feel crushed as mine does at the adjacent field, a desolate brown, the soil vulnerable and denuded of the plants who swaddle, clothe, and feed it.

In a world where mouths of a single species (humans) ingest one out of four nutrients pulled from the soil through a root, there will be some serious consequences to the other 3 million species trying to get by. Even if we are extremely self-interested, our own existence depends on us being able to continue to grow plants to eat in an increasingly unpredictable and hostile climate and to do so in a way that doesn't continue to degrade our literal foundation: the soil.

There's no guarantee, but I believe our best hope will grow from roots like those I'm sampling: the perennial grass *Thinopyrum intermedium* undergoing domestication for grain production. Roots are integral to a plant's life—they are its main source of water, nutrients, and connection—but also to soil formation and the global carbon cycle.

Roots are the main way that carbon enters the soil, feeding the billions of organisms that dwell on and between the smallest clay particles, grains of sand. While they are alive, roots exude a buffet of sugars to entice their microbial partners to support their hunger for phosphorus, nitrogen, and other essential nutrients. They may secrete acids and enzymes akin to our saliva to begin the process of digestion so that nutrients can be absorbed. When roots, their secretions, microbes, and clay particles aggregate together, organic matter is formed. Organic matter accounts for only a small portion of the soil. But as a habitat, food source, and reservoir of water and nutrients, organic matter brings soil functions to life. And since carbon is the main ingredient in organic matter (about 58 percent), to increase soil organic matter is to increase soil carbon.

We know that most of the carbon in the soil came from a root, but there are still many things we do not know. How much carbon, for example, is something we mostly guess about. How many roots die each year versus grow anew? How can we distinguish roots of different species? How deep do roots grow, and how does this vary across soil types and plant communities?

We're out here today, in part, to learn whether the greater root growth of this perennial intermediate wheatgrass (Kernza) has increased the amount of carbon in this soil compared with an adjacent annually cropped field. There are only two ways to increase soil carbon: Add more or lose less. Since perennials have three to fifteen times more roots than their annual counterparts, we're guessing this Kernza field is slowly adding carbon inputs. But this perennial crop can also reduce how often and how long soil is tilled, exposed, and starved of living roots, so it might stem carbon losses too. No one has measured this on more than a couple of farmers' fields so far, so here we are—dodging rocks and hauling roots and soil back to the lab.

This is just one day in what will be many years of work. Eventually, we'll come up with a number that captures how much carbon and nitrogen each sample contains. We'll compare that number with the numbers we got the last time we sampled this field five years ago, and with the annually farmed field across the road, to see whether there are any differences. We'll look at these differences for many fields—as many as we can sample—to seek a pattern. Do the soils under perennial grains have more carbon than the annual grain fields? How does this number change over time? Our initial data are promising—the perennial grain fields show signs of increased soil carbon—but we need more evidence to answer our questions with confidence. We need more samples, more time, more fields planted with perennial grains.

When my daughter was nursing, I was fascinated and awed by how every atom of her body (save some of the oxygen) passed first through mine. There she was, her own person, but made from what had first been part of me. In the same way, we are all children of the soil and the plants who nourish both soils and us. Our relationship with soils hasn't always been healthy, though, and our reliance on annual grains is part of the problem. Perennial grains give us a chance to seek a dynamic that is more exchange than extraction. Not out of some noble sense of conservation but from our animal compunction to survive. Despite the hole we've eaten our way into, perennial grains are our way of rooting for ourselves.

Acknowledgments

Several years ago, before we gave a virtual talk together, a mutual friend and colleague texted us that they were pregaming for our presentation by listening to The Highwomen, a band of singers who joined forces to blend harmonies. At the time, we had met in person and enjoyed energizing conversations. But our colleague's note was a clue to the way our own collaboration and friendship would grow, including through this co-created book.

Like The Highwomen, we want "a house with a crowded table, and a place by the fire for everyone." We want to nurture an Earth crowded with the bustle of life. We are so grateful to everyone who's here, all the creatures and forces at work in the perennial movement.

Thank you, thank you to:

The supergroup of *Living Roots* contributors, for sharing their stories and poems and art; for the discussions, drafts, and final polishes; for the generous trust placed in us as editors and you all as readers.

North American grasslands and coastlines and farms and research plots—especially Konza Prairie, Wauhob Prairie, Marty Bender Nature

Area, North Campus Open Space, and Mount Jumbo—and all the perennial plants we wish we could unfurl here, name by precious name, for their brilliance and companionship.

Tim Crews, for asking Liz when she was going to write a book about perennials, for being the first coworker and most constant mentor with whom Aubrey talked through the project, and for his scientific review and fact-checking. Plus, for inviting us to gaze at the stars and moon.

The Land Institute and the University of California Santa Barbara, for supporting us with time, resources, and intellectual camaraderie. Special thanks to the NoVo Foundation for vital, steady support to The Land Institute that made possible Aubrey's work on this project and to the Regents Humanities Faculty Fellowship at the University of California Santa Barbara for supporting a summer salary for Liz to work on this project.

Rachel Stroer, Terry Evans, Tammy Kimbler, Ricardo Salvador, Mary Pipher, Tessa Peters, Brandon Schlautman, and many more friends and coworkers for encouraging conversations at key points in the process of developing and realizing this book.

Wes Jackson, co-founder of The Land Institute and author of *New Roots for Agriculture*, and Bill Vitek, director of the New Perennials Project and editor of *The Perennial Turn*, for philosophical roots and questions that reach beyond the available answers.

Members of The Land Institute's Perennial Cultures Lab, past and present, especially Lydia Nicholson, for helping envision educational connections; Reece Knapic, for stewarding data systems; and Amy June Breesman, for brainstorming with us about art for the book.

Tom Lynch, for insights about literary editing and the models offered in his collaborative and narrative ecocritical scholarship. Fran Kaye, for teachings about the sufficiency of the Great Plains.

Mandy Brown, for coaching phone calls that worked like punctuation—commas, spaces, exclamation points, and question marks—shaping the flow of creative practice.

The editors and writers of *All We Can Save* and *Black Earth Wisdom* and *Not Too Late*, for collections that continue to inspire us.

Linda Lewis, Kristin Van Tassel, and Greg LeGault, along with Majkin Holmquist and Cody Whetstone, for feedback on introductory drafts and years of writing group conversations with Aubrey.

The Land Institute, including participants in past Ecosphere Studies gatherings, the Association for the Study of Literature & Environment, and the Environmental Studies Program at the University of California Santa Barbara, for providing professional communities and spaces for deep inquiry.

Jessica Papin, for believing in us, providing guidance in the critical early stages, and finding this book a home at Island Press.

Emily Turner, for astute care and partnership in editing, and everyone at Island Press for bringing this book to fruition.

Patrick Archie, for nurturing this and every dream.

Adam Krug, for every step of the way, always, and Bob Krug, for being you. And Hobbes and Bidart for their presence.

Our parents, siblings, and beloved kin of all kinds, far and wide. For being the people around the table, the people in the fields, the people doing the dishes and tending the fires. Those who make and carry the world, now and to come.

Notes

Introduction

"currently responsible for a third of global greenhouse gas emissions" Rob Jackson, *Into the Clear Blue Sky: The Path to Restoring Our Atmosphere* (Scribner, 2024).

"tended these prolific food forests with fire" Gerald Clarke, "Bringing the Past to the Present: Traditional Indigenous Farming in Southern California," in *Indigenous Food Sovereignty in the United States*, ed. Devon A. Mihesuah and Elizabeth Hoover (University of Oklahoma Press, 2019).

"Prairies fed bison, elk, rabbits, prairie chickens, and the many people" Frances W. Kaye, *Goodlands: A Meditation and History on the Great Plains* (AU Press, 2011); Melvin R. Gilmore, *Uses of Plants by the Indians of the Missouri River Region* (University of Nebraska Press, 1991).

"allocate a quarter to a third of their solar harvest" Timothy E. Crews, Wim Carton, and Lennart Olsson, "Is the Future of Agriculture Perennial? Imperatives and Opportunities to Reinvent Agriculture by Shifting from Annual Monocultures to Perennial Polycultures," *Global Sustainability* 1 (2018): e11, https://doi.org/10.1017/sus.2018.11; Alberto Canarini et al., "Root Exudation of Primary Metabolites: Mechanisms and Their Roles in Plant Responses to Environmental Stimuli," *Frontiers in Plant Science* 10 (2019), https://doi.org/10.3389/fpls.2019.00157.

"allocate 50 to 67 percent of fixed carbon" Crews et al., "Is the Future of Agriculture Perennial?"

"three to four years' worth of worldwide emissions from fossil fuels and cement" Ibid.

"300 times the warming power" Stephen J. Del Grosso, Tom Wirth, Stephen M. Ogle, and William J. Parton, "Estimating Agricultural Nitrous Oxide Emissions," *EOS, Transactions American Geophysical Union* 89, no. 51 (2008): 529, https://doi.org/10.1029/2008EO510001.

"Some 55 percent of it comes from agriculture" Kevin J. Wolz, Bruce E. Branham, and Evan H. DeLucia, "Reduced Nitrogen Losses After Conversion of Row Crop Agriculture to Alley Cropping with Mixed Fruit and Nut Trees," *Agriculture, Ecosystems and Environment* 258 (2018): 172–81, https://doi.org/10.1016/j.agee.2018.02.024.   .

"largest source of water pollution" Jennifer Blesh and Laurie E. Drinkwater, "The Impact of Nitrogen Source and Crop Rotation on Nitrogen Mass Balances in the Mississippi River Basin," *Ecological Applications* 23, no. 5 (2013): 1017–35, https://doi.org/10.1890/12-0132.1.

"scavenge much of the nitrogen" Crews et al., "Is the Future of Agriculture Perennial?"

"Eighty thousand years ago" Yan Wu, Dawei Tao, Xiujie Wu, Wu Liu, and Yanjun Cai, "Diet of the Earliest Modern Humans in East Asia," *Frontiers in Plant Science* 13 (2022), https://doi.org/10.3389/fpls.2022.989308.

"Sixty-five thousand years ago" Chris Clarkson et al., "Human Occupation of Northern Australia by 65,000 Years Ago," *Nature* 547 (2017): 306–10, https://doi.org/10.1038/nature22968.

"highlighted in *Parable of the Sower*" Maayan Kreitzman, Eric Toensmeier, Kai M. A. Chan, Sean Smukler, and Navin Ramankutty, "Perennial Staple Crops: Yields, Distribution, and Nutrition in the Global Food System," *Frontiers in Sustainable Food Systems*, 4 (2020), https://doi.org/10.3389/fsufs.2020.588988.

"Walnuts and pistachios in Kyrgyzstan" Ibid.

"Camas bulbs in western North America" Sidney Fellows, Georgia Hart-Fredeluces, Nolan Brown, Morey Burnham, and the Language and Culture Preservation Department, "Strengthening Relationships to Traditional Foodways: Adapting Food Practices Through Camas Cultivation Experiments on the Fort Hall Indian Reservation," *Journal of Ethnobiology* 43, no. 2 (2023): 85–100, https://doi.org/10.1177/02780771231176361.

"Protein-rich and drought-tolerant mesquite" Kreitzman et al., "Perennial Staple Crops."

"some 60 to 80 percent of global cropland" Sean R. Asselin, Anita L. Brûlé-Babel, David L. Van Tassel, and Douglas J. Cattani, "Genetic Analysis of Domestication Parallels in Annual and Perennial Sunflowers (*Helianthus* spp.): Routes to Crop Development," *Frontiers in Plant Science* 11 (2020), https://doi.org/10.3389/fpls.2020.00834.

Part I

"more than 30 percent of the planet's land surface" Masumi Hisano, Eric B. Searle, and Han Y. H. Chen, "Biodiversity as a Solution to Mitigate Climate Change Impacts on the Functioning of Forest Ecosystems," *Biological Reviews* 93, no. 1 (2018): 439–56, https://doi.org/10.1111/brv.12351.

"two thirds of all terrestrial plants and animals" Paul Hawken, *Drawdown: The Most Comprehensive Plan Ever Proposed to Reverse Global Warming* (Penguin Books, 2017).

"vast networks of fungal mycelia" Leah Penniman, *Farming While Black* (Chelsea Green, 2018), 80.

"recent pressure from increasing temperatures" Constance I. Millar and Nathan L. Stephenson, "Temperate Forest Health in an Era of Emerging Megadisturbance," *Science* 349, no. 6250 (2015): 823–6, https://doi.org/10.1126/science.aaa9933.

"absorb more than a quarter of human carbon emissions" William R. L. Anderegg et al., "Climate-Driven Risks to the Climate Mitigation Potential of Forests," *Science* 368, no. 6497 (2020), https://doi.org/10.1126/science.aaz7005.

"up to a third of current carbon emissions" Hawken, *Drawdown*.

"cultivating elaborate gardens" Devon G. Peña, *Mexican Americans and the Environment: Tierra y Vida* (University of Arizona Press, 2005), 51; Anabel Ford, "Dominant Plants of the Maya Forest and Gardens of El Pilar: Implications for Paleoenvironmental Reconstructions," *Journal of Ethnobiology* 28, no. 2 (2008): 179–99, https://doi.org/10.2993/0278-0771-28.2.179.

"transforms a rainforest into a food forest" Judith Carney and Richard Nicholas Rosomoff, *In the Shadow of Slavery: Africa's Botanical Legacy in the Atlantic World* (University of California Press, 2009), 25.

"Other agroforestry practices include" "What Is Agroforestry?," US Department of Agriculture National Agroforestry Center, 2012, https://www.fs.usda.gov/nac/assets/documents/workingtrees/infosheets/WhatIs Agroforestry07252014.pdf.

"increased 40 percent" Andrea De Stefano and Michael G. Jacobson, "Soil Carbon Sequestration in Agroforestry Systems: A Meta-analysis," *Agroforestry Systems* 92 (2018): 285–99, https://doi.org/10.1007/s10457-017-0147-9.

"offset 34 percent of US annual emissions from coal, oil, and gas" Shea Swenson, "Is Agroforestry the Key to Hardier Row Crops?," *Modern Farmer*, December 5, 2022, https://modernfarmer.com/2022/12/agroforestry/.

"reduced annual nitrous oxide fluxes by 25 to 83 percent" Kevin J. Wolz, Bruce E. Branham, and Evan H. DeLucia, "Reduced Nitrogen Losses After Conversion of Row Crop Agriculture to Alley Cropping with Mixed Fruit and Nut Trees," *Agriculture, Ecosystems and Environment* 258 (2018): 172–81, https://doi.org/10.1016/j.agee.2018.02.024.

"74 percent lower nitrous oxide emissions" Chukwudi C. Amadi, Ken C. J. Van Rees, and Richard E. Farrell, "Soil–Atmosphere Exchange of Carbon Dioxide, Methane and Nitrous Oxide in Shelterbelts Compared with Adjacent Cropped Fields," *Agriculture, Ecosystems & Environment* 223 (2016): 123–34, https://doi.org/10.1016/j.agee.2016.02.026.

"top twenty climate solutions" Hawken, *Drawdown.*

"Trees stabilize temperatures belowground" Kevin J. Wolz et al., "Frontiers in Alley Cropping: Transformative Solutions for Temperate Agriculture," *Global Change Biology* 24, no. 3 (2018): 883–94, https://doi.org/10.1111/gcb.13986.

"reduce water pollution from fertilizer" Wolz et al., "Reduced Nitrogen Losses."

"provide crucial habitat" Gary Bentrup, Jennifer Hopwood, Nancy Lee Adamson, and Mace Vaughan, "Temperate Agroforestry Systems and Insect Pollinators: A Review," *Forests* 10, no. 11 (2019): 981, https://doi.org/10.3390/f10110981.

"A recent analysis of health benefits" Kathleen L. Wolf, Sharon T. Lam, Jennifer K. McKeen, Gregory R. A. Richardson, Matilda van den Bosch, and Adrina C. Bardekjian, "Urban Trees and Human Health: A Scoping Review," *International Journal of Environmental Research and Public Health* 17, no. 12 (2020): 4371, https://doi.org/10.3390/ijerph17124371.

"nearly three times annual crop income" "The Opportunity in Permanent Crops," Equilibrium Capital, 2013, https://eq-cap.com/wp-content/uploads/2020/05/The_Opportunity_in_Permanent_Crops-2013_09_04.pdf.

"more efficient use of light and space" Wolz et al., "Frontiers in Alley Cropping"; Matthew M. Smith, Gary Bentrup, Todd Kellerman, Katherine MacFarland, Richard Straight, and Lord Ameyaw, "Windbreaks in the United States: A Systematic Review of Producer-Reported Benefits, Challenges, Management Activities and Drivers of Adoption," *Agricultural Systems* 187 (2021), https://doi.org/10.1016/j.agsy.2020.103032.

"2022 passage of historic climate change legislation" Erik Hoffner, "American Agroforestry Accelerates with New Funding Announcements," *Mongabay*, October 31, 2022, https://news.mongabay.com/2022/10/american-agroforestry-accelerates-with-new-funding-announcements/.

"researchers in Missouri recently mapped marginal lands" Bhuwan Thapa, Sarah Lovell, Ronald Revord, Phillip Owens, Michael Gold, and Nicholas Meier, "Significant Opportunities for Tree Crop Expansion on Marginal Lands in the Midwest, USA," *Journal of Sustainable Agriculture and Environment* 2, no. 4 (2023): 412–23.

Throw It Upwards

"cleared virgin old growth forest" Lucile Maertens and Adrienne Stork, "The Real Story of Haiti's Forests," *Books & Ideas*, October 9, 2017, https://booksandideas.net/The-Real-Story-of-Haiti-s-Forests.

"$115 billion" Catherine Porter, Constant Méheut, Matt Apuzzo, and Selam Gebrekidan, "The Root of Haiti's Misery: Reparations to Enslavers," *New York Times*, May 20, 2022, https://www.nytimes.com/2022/05/20/world/americas/haiti-history-colonized-france.html.

"Haitians exported mahogany" M. R. O'Connor, "One of the Most Repeated Facts About Haiti Is a Lie," *Vice*, October 13, 2016, https://www.vice.com/en/article/one-of-the-most-repeated-facts-about-deforestation-in-haiti-is-a-lie/.

"less than 1 percent" S. Blair Hedges, Warren B. Cohen, Joel Timyan, and Zhiqiang Yang, "Haiti's Biodiversity Threatened by Nearly Complete Loss of Primary Forest," *PNAS* 115, no. 46: 11850–5, https://doi.org/10.1073/pnas.1809753115.

"Haiti's forest cover is around 32 percent" O'Connor, "One of the Most Repeated Facts."

"diasporic delegation" Ayiti Resurrect, https://www.ayitiresurrect.org/.

"Mohican homelands" "Brief History," Stockbridge-Munsee Community Band of Mohican Indians, accessed September 4, 2024, https://www.mohican.com/brief-history/.

"Soul Fire Farm" Soul Fire Farm, https://www.soulfirefarm.org/.

"more than 40 inches" Steven E. Hollinger, *Midwestern Climate Center Soils Atlas and Database*, Illinois State Water Survey (Illinois Department of Energy and Natural Resources, 1995).

"ranks agricultural soils" "Ranking of Mineral Soils," N.Y. Comp. Codes R. & Regs. Tit. 1 § 370.8, https://www.law.cornell.edu/regulations/new-york/1-NYCRR-370.8.

"*fanya-juu*" Julianna White, "Terracing Practice Increases Food Security and Mitigates Climate Change in East Africa," *CGIAR Research Program on Climate Change, Agriculture and Food Security*, May 11, 2016, https://ccafs.cgiar.org/news/terracing-practice-increases-food-security-and-mitigates-climate-change-east-africa#.WJTUEhsrLIU.

"between 1,000 and 2,000 rising generation" "2024 Annual Report," Soul Fire Farm, https://www.soulfirefarm.org/.

"Indigenous microorganisms" Baduru Lakshman Kumar and DVR Sai Gopal, "Effective Role of Indigenous Microorganisms for Sustainable Environment," *3 Biotech* 5, no. 6 (2015): 867–76.

"We feel it very important that we obtain HOMES" Joseph R. Johnson to Gen O. O. Howard, August 4, 1865, Unregistered Letters Received, series 457, DC Assistant Commissioner, Bureau of Refugees, Freedmen, & Abandoned Lands, Record Group 105, National Archives, https://www.freedmen.umd.edu/J%20Johnson.htm.

"Tubman planted an apple orchard" Frank C. Drake, "The Moses of Her People," *New York Herald*, September 2, 1907.

"essential laborers on apple orchards" Tick Root, "The Jamaican Apple Pickers of Upstate New York," *The New York Times*, October 5, 2017, https://www.nytimes.com/2017/10/05/nyregion/the-jamaican-apple-pickers-of-upstate-new-york.html.

"bear a particular burden" Leah Penniman, *Farming While Black: Soul Fire Farm's Practical Guide to Liberation on the Land* (Chelsea Green, 2018).

"**sheep fever of the early 1800s**" Marla Miller and Emily Whitted, "Sheep Fever, Fever Break: Sheep in Nineteenth-Century Hampshire County," SustainableEweMass, accessed September 4, 2024, https://www.ewemass.org /sheep-fever-fever-break-sheep-in-nineteenth-century-hampshire-county .html.

The Cactus Forest

"**historically consumed more than 100 native plants**" Gary Paul Nabhan et al., "An Aridamerican Model for Agriculture in a Hotter, Water Scarce World," *Plants People Planet* 2, no. 6 (2020): 627–39, https://doi .org/10.1002/ppp3.10129.

"**In the desert . . . the very fauna and flora proclaim**" Joseph Wood Krutch, *The Desert Year* (William Sloane Associates, 1952), 181–82.

A Food Forest for Southeast Atlanta

"**nearly one in four**" A. Dewey, V. Harris, M. Hake, and E. Engelhard, *Map the Meal Gap 2024: An Analysis of County and Congressional District Food Insecurity and County Food Cost in the United States in 2022* (Feeding America, 2024).

From the Fringe

"**these species began to show their face**" William K. Stevens, *Miracle Under the Oaks: The Revival of Nature in America* (Gallery Books, 1996).

"**making its way into natural openings over time**" For more exploration of creative disturbance, see Robin Wall Kimmerer, *The Serviceberry: An Economy of Gifts and Abundance* (Scribner, 2024).

In the Presence of an Olive Tree

"**cultures have left an infinity of remnants**" Edward Said, "History, Literature, Geography," in *Reflections on Exile and Other Essays* (Harvard University Press, 2000).

"**have remained in our culture for more than 1,500 years**" Omar Tesdell, Yusra Othman, and Saher Alkhoury, "Rainfed Agroecosystem Resilience in the Palestinian West Bank, 1918–2017," *Agroecology and Sustainable Food Systems* 43, no. 1 (2019): 21–39, https://doi.org/10.1080 /21683565.2018.1537324.

"thinking you know and making it possible to be known" I draw on Søren Kierkegaard, "Love's Hidden Life and Its Recognisability by Its Fruits," in *Works of Love*, trans. Howard V. Hong and Edna H. Hong (Harper Perennial Modern Thought, 2009).

"have grown, flowered, died, and given seed each year" Omar Imseeh Tesdell, Amna Othman, Yusra Othman, Yara Dowani, Renad Shqeirat, Yara Bamieh, Elizabeth Tesdell, and Saher Khouri, *Palestinian Wild Food Plants* (Khalil Sakakini Cultural Center, 2018), https://archive.org/details /palwildfoodplants2018.

"We share our work" Omar Tesdell, Yusra Othman, Yara Dowani, Samir Khraishi, Saher Alkhoury, Mary Deeik, Aubrey Streit Krug, Brandon Schlautman, and David Van Tassel, "Envisioning Perennial Agroecosystems in Palestine," *Journal of Arid Environments* 175 (2020): 104085, https: //doi.org/10.1016/j.jaridenv.2019.104085; Tesdell et al., "Rainfed Agro- ecosystem Resilience"; Tesdell et al., *Palestinian Wild Food Plants*.

The ideas for this essay were developed through conversations with many colleagues in Palestine and the United States and emerged first as a speech delivered at the Prairie Festival at the Land Institute in 2022.

Makwa-Miskomin: Bear's Red Berry

"Anishinaabe" *Anishinaabe*, or *Anishinaabeg* in the plural, has several translations. I am using it to refer to the Ojibwe, also spelled *Ojibwa*, and Chippewa, and the words in this essay that are not English are Ojibwe words. There are several tribes related to the Ojibwe who also call them- selves Anishinaabeg.

"do not leave enough" Megan Ulu-Lani Boyanton, "The Complex Case of Growing Native Plants," *High Country News*, April 1, 2024, https: //www.hcn.org/issues/the-complex-case-of-growing-native-plants/.

The Odyssey of Coffee

"used and maintained for centuries" Kristoffer Hylander, Sileshi Nemomissa, Joern Fischer, Beyene Zewdie, Biruk Ayalew, and Ayco JM Tack, "Lessons from Ethiopian Coffee Landscapes for Global Conservation in a Post-Wild World," *Communications Biology* 7, no. 1 (2024): 714.

"Coffeehouses burgeoned across the Near East" Markman Ellis, *The Coffee-House: A Cultural History* (Weidenfeld & Nicolson, 2011).

"Coffeehouses proliferated throughout Europe" Ibid.

"arrived on the scene in the late 1860s" Stuart McCook, *Coffee Is Not Forever: A Global History of the Coffee Leaf Rust* (Ohio University Press, 2019).

"an idea later disproven" Stuart McCook and John Vandermeer, "The Big Rust and the Red Queen: Long-Term Perspectives on Coffee Rust Research," *Phytopathology* 105, no. 9 (2015): 1164–73.

"'technification' of coffee" Robert A. Rice, "A Place Unbecoming: The Coffee Farm of Northern Latin America," *Geographical Review* 89, no. 4 (1999): 554–79.

"discovered 30 ant species and 126 beetle species" Ivette Perfecto and John Vandermeer, *Coffee Agroecology: A New Approach to Understanding Agricultural Biodiversity, Ecosystem Services and Sustainable Development* (Routledge, 2015).

"*Finca Irlanda*" Ivette Perfecto, John Vandermeer, and Stacy M. Philpott, "Complex Ecological Interactions in the Coffee Agroecosystem," *Annual Review of Ecology, Evolution, and Systematics* 45, no. 1 (2014): 137–58.

"Their efforts are beautifully complementary" Lauren Schmitt, Russell Greenberg, Guillermo Ibarra-Núñez, Peter Bichier, Caleb E. Gordon, and Ivette Perfecto, "Cascading Effects of Birds and Bats in a Shaded Coffee Agroforestry System," *Frontiers in Sustainable Food Systems* 5 (2021): 512998; Kimberly Williams-Guillén, Ivette Perfecto, and John Vandermeer, "Bats Limit Insects in a Neotropical Agroforestry System," *Science* 320, no. 5872 (2008): 70.

"enhancing the natural pest control mechanisms" Ivette Perfecto, John Vandermeer, and Angus Wright, *Nature's Matrix: Linking Agriculture, Biodiversity Conservation and Food Sovereignty* (Routledge, 2019).

"effectively protect the beetle that regulates the pest population" John Vandermeer, Ivette Perfecto, and Stacy Philpott, "Ecological Complexity and Pest Control in Organic Coffee Production: Uncovering an Autonomous Ecosystem Service," *BioScience* 60, no. 7 (2010): 527–37.

"sought fair trade certifications" Maria Elena Martinez-Torres, *Organic Coffee: Sustainable Development by Mayan Farmers* (Ohio University Press, 2006).

"harnessed the fair trade coffee market" Lindsay Naylor, *Fair Trade Rebels: Coffee Production and Struggles for Autonomy in Chiapas* (University of Minnesota Press, 2019).

Fruit Finding

"De Soto initiated war" John R. Fordyce, "Explorations of De Soto," *The Military Engineer* 28, no. 158 (1936): 114–9.

"spoke of their value as a crucial food source" William Bartram, *Travels Through North & South Carolina, Georgia, East & West Florida, the Cherokee Country, the Extensive Territories of the Muscogulges or Creek Confederacy, and the Country of the Chactaws* (James & Johnson, 1791).

"Yardley Taylor" Eugene Scheel, "Underground Railroad: Journey to Freedom Was Risky for Slaves and Guides," *The History of Loudoun County, Virginia*, accessed March 4, 2025, https://www.loudounhistory .org/history/underground-railroad/.

"grafting and growing superior fruit and nut cultivars" A. Glenn Crothers, *Quakers Living in the Lion's Mouth: The Society of Friends in Northern Virginia, 1730–1865* (University Press of Florida, 2012).

"one of the most commonly grown trees" S. A. Aminu, Y. Ibrahim, H. A. Ismail, and I. O. Ibrahim, "Medicinal and Traditional Utilization of African Ebony (*Diospyros mespiliformi*): A Review," *International Journal of Current Microbiology and Applied Sciences* 10, no. 06 (2021): 811-7.

"one of the oldest known human foods" National Research Council, *Lost Crops of Africa: Volume III: Fruits* (The National Academies Press, 2008), https://doi.org/10.17226/11879.

"used to produce various medicinals" Michael W. Twitty, *Fighting Old Nep: The Foodways of Enslaved Afro-Marylanders, 1634–1864* (2006).

"adept at transforming traditional European recipes" Kelley Fanto Deetz, "How Enslaved Chefs Helped Shape American Cuisine," *Smithsonian Magazine*, July 20, 2018, https://www.smithsonianmag.com /history/how-enslaved-chefs-helped-shape-american-cuisine-180969697/.

"A $15 prize" G. Marshall, "Fall Nut Show," *The Hoosier Kernel* 4, no. 3 (1957): 7.

"Claypool made 1,500 crosses" William Heiman, Jr., "Notes from the Editor," *The Hoosier Kernel* 37, no. 5 (1990): 14.

Hazel-Bush and Willow

"yellowed pages of old local histories" Newton Bateman, Paul Selby, J. Seymour Currey, and Charles J. Scofield, eds., *Historical Encyclopedia of Illinois and History of Hancock County*, Vol. II. (Munsell Publishing Co., 1921), 1270; Thomas H. Gregg, *History of Hancock County, Illinois* (Charles C. Chapman & Co., 1880), 878.

"conversation with nature" The exact formulation of the "conversation with nature" I quote here is Wes Jackson's, which he presents as a summary of Wendell Berry's ideas. Representative (but not necessarily earliest) citations: Wes Jackson, "The Necessity and Possibility of an Agriculture Where Nature Is the Measure," *Conservation Biology* 22, no. 6 (2008):1376–7; Wendell Berry, "Nature as Measure," in *What Are People For?* (North Point Press, 1990), 209.

Part II

"fourth largest plant family" Matt A. Sanderson, David Wedin, and Ben Tracy, "Grassland: Definition, Origins, Extent, and Future," in *Grassland Quietness and Strength for a New American Agriculture*, ed. Walter F. Wedin and Steven L. Fales (American Society of Agronomy, Crop Science Society of America, and Soil Science Society of America, 2009), 59.

"covered nearly 40 percent" Bradford P. Wilcox, Samuel D. Fuhlendorf, John W. Walker, Dirac Twidwell, X. Ben Wu, Laura E. Goodman, Morgan Treadwell, and Andrew Birt, "Saving Imperiled Grassland Biomes by Recoupling Fire and Grazing: A Case Study from the Great Plains," *Frontiers in Ecology and the Environment* 20, no. 3 (2022): 179–86, https://doi.org/10.1002/fee.2448.

"Hotspots of biodiversity and clean water" Ibid.

"have flourished in grasslands" "Valuing Grasslands: Critical Ecosystems for Nature, Climate, and People," Grasslands, Rangelands, Savannahs and Shrublands Alliance, Conservation International, World Wildlife Fund, Nature Conservancy, Bird Life International, and Plantlife, 2024, https://www.nature.org/content/dam/tnc/nature/en/documents/Valuing-Grasslands-Critical-Ecosystems-for-Nature-Climate-and-People-Discussion-Paper.pdf.

"nearly half of the world's grasslands have been lost" Wilcox et al., "Saving Imperiled Grassland Biomes."

"overcome by nonnative species" Ibid.

"among the most threatened grasslands on Earth" Ibid.

"once covered a billion acres" Matthew C. Reeves, Michael Krebs, Ian Leinwand, David M. Theobald, and John E. Mitchell, "Rangelands on the Edge: Quantifying the Modification, Fragmentation, and Future Residential Development of U.S. Rangelands," General Technical Report RMRS-GTR-382 (US Department of Agriculture, Forest Service, Rocky Mountain Research Station, 2018), 24.

"more than 34 percent of these lands" Ibid.

"96 percent of the tallgrass prairies" Nathan F. Sayre, "A History of North American Rangelands," in *Rangeland Wildlife Ecology and Conservation*, ed. Lance B. McNew, David K. Dahlgren, and Jeffrey L. Beck (Springer, 2003), 51.

"among the most ecologically intact landscapes" Ibid.

"grasslands covered 20 percent of the state before 1850" M. Kat Anderson, *Tending the Wild: Native American Knowledge and the Management of California's Natural Resources* (University of California Press, 2005), 28.

"have traditionally been harvested" Ibid., 29.

"knew it was time to dig edible bulbs" Ibid., 52, 167.

"used sophisticated baskets to collect seeds" Ibid., 52, 129, 165.

"prairie turnip paradox" Lisa M. Castle, "The Prairie Turnip Paradox: Contributions of Population Dynamics, Ethnobotany, and Community Ecology to Understanding *Pediomelum esculentum* Root Harvest on the Great Plains" (PhD diss., University of Kansas., 2006); William M. Crawford, "Daƙod Wóokaȟniġe Ohna Ṫípsiŋna Awaŋyakapi (Protecting Prairie Turnips Through Traditional Dakota Wisdom and Understanding)" (Master's thesis, University of Colorado Boulder, 2023); Kelly Kindscher, *Edible Wild Plants of the Prairie: An Ethnobotanical Guide* (University Press of Kansas, 2024); April Stahnke, Michelle Hayes, Karen Meyer, Karla Witt, Jeanna Weideman, Anne P. Fernando, Rhoda Burrows, and R. Neil Reese, "Prairie Turnip *Pediomelum esculentum* (Pursh) Rydb.: Historical and Modern Use, Propagation, and Management of a New Crop," *Native Plants Journal* 9, no. 1 (2008): 46–58, https://doi.org/10.2979/NPJ.2008.9.1.46.

"tended grasslands with carefully timed fires" Anderson, *Tending the Wild*; Wilcox et al., "Saving Imperiled Grassland Biomes."

"sustain the ones who sustain you" Robin Wall Kimmerer, *Braiding Sweetgrass: Indigenous Wisdom, Scientific Knowledge, and the Teachings of Plants* (Milkweed Editions, 2013), 183

"the grass has time to grow" Robin Wall Kimmerer, *Braiding Sweetgrass: Indigenous Wisdom, Scientific Knowledge, and the Teachings of Plants* (Milkweed Editions, 2013), 164.

"breaking up large corn and soy plantings with prairie strips" "Science-Based Trials of Rowcrops Integrated with Prairie Strips," Iowa State University, 2025, https://www.nrem.iastate.edu/research/STRIPS/.

"remarkably efficient at scavenging nutrients" Timothy E. Crews, Wim Carton, and Lennart Olsson, "Is the Future of Agriculture Perennial? Imperatives and Opportunities to Reinvent Agriculture by Shifting from Annual Monocultures to Perennial Polycultures," *Global Sustainability* 1 (2018): e11, https://doi.org/10.1017/sus.2018.11.

"lost eight times more nitrogen and 35 times more sediment" Heidi Asbjornsen et al., "Targeting Perennial Vegetation in Agricultural Landscapes for Enhancing Ecosystem Services," *Renewable Agriculture and Food Systems* 29, no. 2 (2014): 101–25, https://doi.org/10.1017/S1742170512000385.

"reduced nitrogen loss by 68 percent" Ranjith P. Udawatta, Harold E. Garrett, and Robert Kallenbach, "Agroforestry Buffers for Nonpoint Source Pollution Reductions from Agricultural Watersheds," *Journal of Environmental Quality* 40, no. 3 (2011): 800–6, https://doi.org/10.2134/jeq2010.0168.

"grow so robustly without fertilizer" Crews et al., "Is the Future of Agriculture Perennial?"

"Evolved to thrive in dry places" Asbjornsen et al., "Targeting Perennial Vegetation."

"ecological plasticity" Ibid.

"preferred form of perennial plant cover for sequestering carbon in California" Pawlok Dass, Benjamin Z. Houlton, Yingping Wang, and David Warlind, "Grasslands May Be More Reliable Carbon Sinks than Forests in California," *Environmental Research Letters* 13, (2018): 074027, https://doi.org/10.1088/1748-9326/aacb39.

"protecting old-growth remnants" Elise Buisson, Sally Archibald, Alessandra Fidelis, and Katharine N. Suding, "Ancient Grasslands Guide Ambitious Goals in Grassland Restoration," *Science* 377, no. 6606 (2022): 594–8, https://doi.org/10.1126/science.abo4605.

Landscapes of Abundance

"the story of the Iniskim" Clark Wissler, *Ceremonial Bundles of the Blackfoot Indians* (The Trustees, 1912), 242.

"define Indigenous lands as 'wild' and 'empty'" Wilderness Act of 1964, Pub. L. 88–577.

"the serviceberry was the most important vegetable food" Clark Wissler, *Material Culture of the Blackfoot Indians* (The Trustees, 1910), 20.

"combination of drying or freezing them" Just as a note, freezing fresh berries creates a different consistency when they are thawed, and many of the old recipes used for dry berries cannot be used with frozen berries.

"This will enrich the soil" Wissler, *Ceremonial Bundles*, 200.

One Prairie Strip at a Time

"reducing water runoff by 37 percent" Lisa A. Schulte, Jarad Niemi, Matthew J. Helmers, Matt Liebman, J. Gordon Arbuckle, David E. James, Randall K. Kolka et al., "Prairie Strips Improve Biodiversity and the Delivery of Multiple Ecosystem Services from Corn–Soybean Croplands," *Proceedings of the National Academy of Sciences* 114, no. 42 (2017): 11247–52.

"twenty-three more carabid species" Andres Vargas, "Ground Beetle Response to Prairie Strips and Their Potential Ecosystem Service Delivery in Crop Fields" (M.S. thesis, Iowa State University, 2024).

"twice as many birds on fields with prairie strips" Jordan C. Giese, Lisa A. Schulte, and Robert W. Klaver, "Bird Community Response to Field-Level Integration of Prairie Strips," *Agriculture, Ecosystems & Environment* 374 (2024): 109075.

"improve water infiltration and thus reduce runoff" Eric J. Henning, Randall K. Kolka, and Matthew J. Helmers, "Infiltration Within Native Prairie Vegetative Strips Embedded in Row Crop Fields Across Iowa," *Journal of Soil and Water Conservation* 79, no. 1 (2024): 43–52.

"90 percent less total suspended sediment" Jessica A. Stephenson, Matt Liebman, Jarad Niemi, Richard Cruse, John Tyndall, Chris Witte, Dave James, and Matthew Helmers, "The Influence of Prairie Strips Sown in Midwestern Corn and Soybean Fields on Sediment Discharge Throughout the Year," *Journal of Soil and Water Conservation* 79, no. 2 (2024): 87–98; Matthew Helmers, Iowa State University, unpublished data.

"needed to get more serious about our conservation practices" Katrina Markel, "'The Prairie Sells Itself': Iowa Farmers Discover Advantages to Planting Prairie Strips," *KMTV Omaha*, May 23, 2024, https://www.3newsnow.com/southwest-iowa/the-prairie-sells-itself-iowa-farmers-discover-advantages-to-planting-prairie-strips.

"This water quality stuff is real" "Eric Hoien—Landowner," Testimonials, Science-Based Trials of Rowcrops Integrated with Prairie Strips, Iowa State University, accessed April 6, 2025, https://www.nrem .iastate.edu/research/STRIPS/testimonials/eric-hoien-landowner.

"The only bottom line is not the dollar bill" "Prairie Strips—Bringing Back the Edges," Practical Farmers of Iowa, October 5, 2021, 6 min., 11 sec., https://www.youtube.com/watch?v=NSCxiIQ7mI8.

"improving the microbial community in the soil" "Prairie Strips Help Farmers, Land, Water," *Working for Clean Water: 2014 Iowa Watershed Successes*, Iowa Department of Natural Resources, 2014, 11, https: //www.iowadnr.gov/environmental-protection/water-quality/watershed -improvement/watershed-success.

"more wildness in my landscape" "Prairie Strips—Bringing Back the Edges."

"a greater proportion of farmers have indicated a willingness" J. Gordon Arbuckle, *2024 Summary Report—Iowa Farm and Rural Life Poll*, SOC 3111 (Iowa State University Extension and Outreach, 2024).

"more than 6,500 acres of prairie strips in Iowa" Cathie Feather, *Conservation Reserve Program Monthly Summary—November 2024* (US Department of Agriculture Farm Service Agency, 2024).

From Corn Belt to Pasture

"carved out of native prairie" "The Iron Horse of the Prairie State," Illinois History & Lincoln Collections, University of Illinois Library History Collection, accessed July 28, 2024, https://publish.illinois.edu /ihlc-blog/exhibits/the-iron-horse-of-the-prairie-state/.

"this quarter section was a mosaic" "Wisconsin Public Land Survey Records: Original Field Notes and Plat Maps," Wisconsin Board of Commissioners of Public Lands, accessed January 17, 2025, https://digicoll .library.wisc.edu/SurveyNotes/.

"80 percent of the cropland is corn or soybeans" "Census of Agriculture," United States Department of Agriculture, https://www.nass .usda.gov/AgCensus/.

"the reach of a plant's root system can be more than doubled" David H. McNear Jr., "The Rhizosphere—Roots, Soil and Everything In Between," *Nature Education Knowledge* 4, no. 3 (2013): 1.

"have steadily lost soil carbon since the beginning" Clarissa L. Dietz, Randall D. Jackson, Matthew D. Ruark, and Gregg R. Sanford, "Soil Carbon Maintained by Perennial Grasslands over 30 Years but Lost in Field Crop Systems in a Temperate Mollisol," *Communications Earth & Environment* 5, no. 1 (2024): 360.

"provide habitat for an estimated 28.6 million pairs of grassland birds" Laura K. Paine, Randy D. Jackson, Zach Raff, Eric G. Booth, Claudio Gratton, Abraham Gibson, David LeZaks, Sarah Lloyd, and Carl Wepking, "Well-managed Perennial Pasture: Setting the Gold Standard for Ecosystem Services," Grassland 2.0, November, 2021, https://grasslandag .org/wp-content/uploads/sites/391/2021/12/FINAL-Ecosystem-services -fact-sheet.pdf.

"more than 75 percent of grass-fed beef sold in the United States is imported" Renee Cheung and Paul McMahon, *Back to Grass: The Market Potential for U.S. Grassfed Beef* (The Stone Barns Center for Food and Agriculture, 2017).

Root Foods

Information in this chapter comes from Kelly Kindscher, *Edible Wild Plants of the Prairie: An Ethnobotanical Guide* (University Press of Kansas, 2024); Kelly Kindscher, Loren Yellow Bird, Michael Yellow Bird, and Logan Sutton, *Sahnish (Arikara) Ethnobotany* (Society of Ethnobiology, 2021).

The Beauty of Polycultures

"provides habitat to 4,000 native plant species" Pablo Modernel, Walter A. H. Rossing, Marc Corbeels, Santiago Dogliotti, Valentín Picasso, and Pablo Tittonell, "Land Use Change and Ecosystem Service Provision in Pampas and Campos Grasslands of Southern South America," *Environmental Research Letters* 11, no. 11 (2016): 113002, https: //iopscience.iop.org/article/10.1088/1748-9326/11/11/113002.

"one of the most threatened biomes in the world" Santiago Baeza and José M. Paruelo, "Land Use/Land Cover Change (2000–2014) in the Rio de la Plata Grasslands: An Analysis Based on MODIS NDVI Time Series," *Remote Sensing* 12, no. 3 (2020): 381, https://doi.org/10.3390 /rs12030381.

"Optimal grazing management can preserve grassland diversity" Soledad Orcasberro, Laura Astigarraga, Marta Moura Kohmann, Pablo Modernel, and Valentín D. Picasso, "Ecological Intensification in Grasslands for Resilience and Ecosystem Services: The Case of Beef Production Systems on the Campos Grasslands of South America," in *Creating Resilient Landscapes in an Era of Climate Change*, ed. Amin Rastandeh and Meghann Jarchow (Routledge, 2022), 94–115.

"famous report from the United Nations Food and Agriculture Organization" Henning Steinfeld, Pierre Gerber, Tom D. Wassenaar, Vincent Castel, and Cees De Haan, *Livestock's Long Shadow: Environmental Issues and Options* (Food and Agriculture Organization of the United Nations, 2006), https://openknowledge.fao.org/server/api/core/bitstreams/36ade937-4641-46ed-aac4-6162717d8a7f/content.

"established on 90 percent of the country's agricultural land" Fernando García-Préchac, Oswaldo Ernst, Guillermo Siri-Prieto, and José A. Terra, "Integrating No-Till into Crop–Pasture Rotations in Uruguay," *Soil and Tillage Research* 77, no. 1 (2004): 1–13, https://doi.org/10.1016/j.still.2003.12.002.

"About 1 million hectares of grassland was converted" Jimena Perez Rocha, *El Estado del Campo Natural en el Uruguay* (Food and Agriculture Organization of the United Nations, 2020), https://doi.org/10.4060/cb0989es.

"feedlots have three negative environmental impacts" Pablo Modernel, Laura Astigarraga, and Valentín Picasso, "Global Versus Local Environmental Impacts of Grazing and Confined Beef Production Systems," *Environmental Research Letters* 8, no. 3 (2013): 035052.

"grazing systems could potentially help recapture some carbon" Valentín D. Picasso, Pablo D. Modernel, Gonzalo Becoña, Lucía Salvo, Lucia Gutierrez, and Laura Astigarraga, "Sustainability of Meat Production Beyond Carbon Footprint: A Synthesis of Case Studies from Grazing Systems in Uruguay," *Meat Science* 98, no. 3 (2014): 346–54, https://doi.org/10.1016/j.meatsci.2014.07.005.

Part III

"can be conveniently stored, traded, and transported" David Van Tassel and Duane Schrag, "Grains," in *Berkshire Encyclopedia of Sustainability*, ed. Ray C. Anderson (Berkshire Publishing Group, 2010).

"more than half the calories humans eat directly" Joseph M. Awika, "Major Cereal Grains Production and Use Around the World," in *Advances in Cereal Science: Implications to Food Processing and Health Promotion*, ed. Joseph M. Awika, Vieno Piironen, and Scott Bean (ACS Symposium Series, vol. 1089, 2011); Danny Hunter et al., "The Potential of Neglected and Underutilized Species for Improving Diets and Nutrition," *Planta* 250 (2019): 709–29, https://doi.org/10.1007/s00425-019-03169-4; Peiman Milani et al., "The Whole Grain Manifesto: From Green Revolution to Grain Evolution," *Global Food Security* 34 (2022): 100649, https://doi.org/10.1016/j.gfs.2022.100649.

"for 70 percent or more of the calories humans consume" Stuart L. Pimm, *The World According to Pimm: A Scientist Audits the Earth* (McGraw-Hill, 2001).

"Pulses are second only to cereals" Narpinder Singh, "Pulses: An Overview," *Journal of Food Science and Technology* 54 (2017): 853–7, https://doi.org/10.1007/s13197-017-2537-4; John Mendendorp, David DeYoung, Deepa G. Thiagarajan, Randy Duckworth, and Barry Pittendrigh, "A Systems Perspective of the Role of Dry Beans and Pulses in the Future of Global Food Security," in *Dry Beans and Pulses: Production, Processing, and Nutrition* (2nd ed.), ed. Muhammad Siddiq and Mark A. Uebersax (Wiley, 2021).

"increased in global production since 1961" Navin Ramankutty, Zia Mehrabi, Katharina Waha, Larissa Jarvis, Claire Kremen, Mario Herrero, and Loren H. Rieseberg, "Trends in Global Agricultural Land Use: Implications for Environmental Health and Food Security," *Annual Review of Plant Biology* 69 (2018): 789–815, https://doi.org/10.1146/annurev-arplant-042817-040256.

"challenged by stagnating yields and lack of diversity" Deepak K. Ray, Lindsey L. Sloat, Andrea S. Garcia, Kyle F. David, Tariq Ali, and Wei Xie, "Crop Harvests for Direct Food Use Insufficient to Meet the UN's Food Security Goal," *Nature Food* 3 (2022): 367–74, https://doi.org/10.1038/s43016-022-00504-z; Toshichika Iizumi et al., "Historical Changes in Global Yields: Major Cereal and Legume Crops from 1982 to 2006," *Global Ecology and Biogeography* 23, no. 3 (2014): 346–57, https://doi.org/10.1111/geb.12120.

"sustainable, resilient, nutritious, or equitable food systems we need" Aubrey Streit Krug, Emily B. M. Drummond, David L. Van Tassel, and Emily J. Warschefsky, "The Next Era of Crop Domestication Starts Now," *Proceedings of the National Academy of Sciences* 120, no. 14 (2023), https://doi.org/10.1073/pnas.2205769120.

"two thirds of the world's arable landscapes" Jerry D. Glover et al., "Increased Food and Ecosystem Security via Perennial Grains," *Science* 328, no. 5986 (2010): 1638–9, https://doi.org/10.1126/science.1188761.

"comes especially from fossil fuels" Smil, Pimentel, and Malm citations in Timothy E. Crews, Wim Carton, and Lennart Olsson, "Is the Future of Agriculture Perennial? Imperatives and Opportunities to Reinvent Agriculture by Shifting from Annual Monocultures to Perennial Polycultures," *Global Sustainability* 1 (2018): e11, https://doi.org/10.1017/sus.2018.11.

"expect that perennial grains will restore ecological functions" Crews et al., "Is the Future of Agriculture Perennial?"; Lee R. DeHaan and David L. Van Tassel, "Gourmet Grasslands: Harvesting a Perennial Future," *One Earth* 5, no. 1 (2022): 14–17, https://doi.org/10.1016/j.oneear.2021.12.012; Timothy E. Crews et al., "Going Where No Grains Have Gone Before: From Early to Mid-Succession," *Agriculture, Ecosystems & Environment* 223 (2016): 223–38, https://doi.org/10.1016/j.agee.2016.03.012; Timothy E. Crews and Brian E. Rumsey, "What Agriculture Can Learn from Native Ecosystems in Building Soil Organic Matter: A Review," *Sustainability* 9, no. 4 (2017): 578, https://doi.org/10.3390/su9040578.

"perennial grains are an increasingly possible reality" Lee R. DeHaan et al., "Discussion: Prioritize Perennial Grain Development for Sustainable Food Production and Environmental Benefits," *Science of the Total Environment* 895 (2023), 164975, https://doi.org/10.1016/j.scitotenv.2023.164975; Wes Jackson, *New Roots for Agriculture* (University of Nebraska Press, 1980).

"development of perennial rice" Shilai Zhang et al., "Sustained Productivity and Agronomic Potential of Perennial Rice," *Nature Sustainability* 6 (2023): 28–38, https://doi.org/10.1038/s41893-022-00997-3.

"cycles of selection lead plant populations to become crops" David L. Van Tassel, Omar Tesdell, Brandon Schlautman, Matthew J. Rubin, Lee R. DeHaan, Timothy E. Crews, and Aubrey Streit Krug, "New Food Crop Domestication in the Age of Gene Editing: Genetic, Agronomic and Cultural Change Remain Co-Evolutionarily Entangled," *Frontiers in Plant Science* 11 (2020): 789, https://doi.org/10.3389/fpls.2020.00789.

"pathway being taken with Kernza® perennial grain" Prabin Bajgain et al., "Breeding Intermediate Wheatgrass for Grain Production," *Plant Breeding Reviews* 46 (2022): 119–217, https://doi.org/10.1002/9781119874157.ch3.

"mutually beneficial systems of relationship" Aubrey Streit Krug and Omar Imseeh Tesdell, "A Social Perennial Vision: Transdisciplinary Inquiry for the Future of Diverse, Perennial Grain Agriculture," *Plants, People, Planet* 3, no. 4 (2021): 355–62, https://doi.org/10.1002/ppp3.10175.

Toward Perennial Sorghum

"stabilizing dunes and protecting shorelines" Pheonah Nabukalu and Carrie A. Knott, "Development and Evaluation of Methods to Identify Sea Oats Breeding Lines for Beaches with Shallow Dunes," *HortTechnology* 24, no. 4 (2014): 484–9.

"a major food source in parts of Africa and Asia" Md. Saddam Hossain, Md. Nahidul Islam, Md. Mamunur Rahman, Mohammad Golam Mostofa, and Md. Arifur Rahman Khan, "Sorghum: A Prospective Crop for Climatic Vulnerability, Food and Nutritional Security," *Journal of Agriculture and Food Research* 8 (2022): 100300.

"displayed many traits similar to cultivated annual sorghum" Stan Cox, Pheonah Nabukalu, Andrew H. Paterson, Wenqian Kong, Susan Auckland, Lisa Rainville, Sheila Cox, and Shuwen Wang, "High Proportion of Diploid Hybrids Produced by Interspecific Diploid × Tetraploid Sorghum Hybridization," *Genetic Resources and Crop Evolution* 65 (2018): 387–90.

Shallow Roots Run Deep

"only 6 percent of plant species are annuals" Tyler Poppenwimer, Itay Mayrose, and Niv DeMalach, "Revising the Global Biogeography of Annual and Perennial Plants," *Nature* 624 (2023): 109–14, https://doi.org/10.1038/s41586-023-06644-x.

"drew attention to the perennial blind spot" Wes Jackson, *New Roots for Agriculture* (University of Nebraska Press, 1980).

"results were anything but ho-hum" Jerry D. Glover et al., "Harvested Perennial Grasslands Provide Ecological Benchmarks for Agricultural Sustainability," *Agriculture, Ecosystems & Environment* 137 (2010): 3–12, https://doi.org/10.1016/j.agee.2009.11.001.

"yielded eight harvests in a row" Shilai Zhang et al., "Sustained Productivity and Agronomic Potential of Perennial Rice," *Nature Sustainability* 6 (2023): 28–38, https://doi.org/10.1038/s41893-022-00997-3.

Rooting for Ourselves

"perennial grains we're studying are recent developments" David L. Van Tassel, Lee R. DeHaan, and Thomas S. Cox, "Missing Domesticated Plant Forms: Can Artificial Selection Fill the Gap?," *Evolutionary Applications* 3, no. 5–6 (2010): 434–52, https://doi.org/10.1111/j.1752 -4571.2010.00132.x; T. S. Cox et al., "Breeding Perennial Grain Crops," *Critical Reviews in Plant Sciences* 21, no. 2 (March 1, 2002): 59–91, https://doi.org/10.1080/0735-260291044188.

"more than five times that volume of soil into the ocean" David R. Montgomery, "Soil Erosion and Agricultural Sustainability," *Proceedings of the National Academy of Sciences* 104, no. 33 (August 14, 2007): 13268– 72, https://doi.org/10.1073/pnas.0611508104.

"one out of four nutrients pulled from the soil" Vaclav Smil, *Harvesting the Biosphere: What We Have Taken from Nature* (MIT Press, 2015), https: //mitpress.mit.edu/9780262528276/harvesting-the-biosphere/; David G. Jenkins et al., "Global Human 'Predation' on Plant Growth and Biomass," *Global Ecology and Biogeography* 29, no. 6 (2020): 1052–64, https://doi .org/10.1111/geb.13087.

"the other 3 million species" Estimates of Earth's species range from 3 to 100 million according to the National Wildlife Federation, "How Many Species Exist?," 1998, https://www.nwf.org/Home/Magazines /National-Wildlife/1999/How-Many-Species-Exist.

"grow plants to eat in an increasingly unpredictable and hostile climate" The Millennium Assessment has an in-depth summary of ecosystem effects of human activity (https://www.millenniumassessment.org/en /index.html). There are probably countless sources to read on the topic, and here are a few: John Gowdy, "Our Hunter–Gatherer Future: Climate Change, Agriculture and Uncivilization," *Futures* 115 (January 1, 2020): 102488, https://doi.org/10.1016/j.futures.2019.102488; Jonas Jaegermeyr et al., "Climate Change Signal in Global Agriculture Emerges Earlier in New Generation of Climate and Crop Models," August 12, 2021, https: //doi.org/10.21203/rs.3.rs-101657/v1; IPCC, "IPCC Special Report on Climate Change, Desertification, Land Degradation, Sustainable Land Management, Food Security, and Greenhouse Gas Fluxes in Terrestrial Ecosystems," 2019, https://www.ipcc.ch/report/SRCCL/.

"most of the carbon in the soil came from a root" Daniel P. Rasse, Cornelia Rumpel, and Marie-France Dignac, "Is Soil Carbon Mostly Root Carbon? Mechanisms for a Specific Stabilisation," *Plant and Soil* 269, no. 1 (February 1, 2005): 341–56, https://doi.org/10.1007/s11104-004-0907-y.

"three to fifteen times more roots" Andrew R. McGowan et al., "Soil Organic Carbon, Aggregation, and Microbial Community Structure in Annual and Perennial Biofuel Crops," *Agronomy Journal* 111, no. 1 (2019): 128–42, https://doi.org/10.2134/agronj2018.04.0284; Stella Woeltjen, Jessica Gutknecht, and Jacob Jungers, "Age-Related Changes in Root Dynamics of a Novel Perennial Grain Crop," *Grassland Research* 3, no. 1 (2024): 57–68, https://doi.org/10.1002/glr2.12068; Christine D. Sprunger et al., "How Does Nitrogen and Perenniality Influence Belowground Biomass and Nitrogen Use Efficiency in Small Grain Cereals?," *Crop Science* 58, no. 5 (September 2018): 2110–20, https://doi.org/10.2135/cropsci2018.02.0123.

"show signs of increased soil carbon" Laura K. van der Pol et al., "Perennial Grain Kernza® Fields Have Higher Particulate Organic Carbon at Depth than Annual Grain Fields," *Canadian Journal of Soil Science*, September 23, 2022, cjss-2022-0026, https://doi.org/10.1139/cjss-2022-0026.

Contributors

Jacob Birch is a Gamilaraay man living and working on the east coast of Australia. The Gamilaraay people are one of the many distinct nations that make up Australia's Indigenous peoples. Jacob is an academic, entrepreneur, and advocate for food sovereignty and the re-Indigenization of the food and agriculture system. His passion for Australian native grains, and all of the intrinsic benefits they carry, has led him to work across disparate spaces as he takes a systems approach to reawaken native grain foodways and, alongside other Gamilaraay people, leverage that reawakening as a nation-building opportunity.

Kelsi E. Bowens is an Atlanta native who has worked in the environmental equity sector through various organizations in Atlanta, Georgia, for nearly a decade, strategizing solutions for social justice issues at the intersection of environmental challenges. At the time of publication, Kelsi worked as Multisolving Institute's Engagement Manager. Before joining Multisolving Institute, Kelsi worked with The Conservation Fund, where she helped manage and expand urban conservation initiatives—including the Urban Food Forest at Browns Mill—across the United States. Kelsi holds an MBA from Mercer University and a BA in journalism with a minor in environmental law from the University of Georgia. She adores her maltipoo, Grady, and traveling to learn about cultures around the world.

Tim Crews is the chief scientist and director of the International Initiative of The Land Institute in Salina, Kansas. He is currently working to invite and help support researchers from around the world to join in a global expansion of perennial grains development, cropping systems, and social adoption. Tim's own research interests focus on ecological intensification of perennial agroecosystems. He loves exploring the night sky, playing and listening to music, hiking, biking, and managing a prairie restoration cemetery with his wife, Sarah.

Colin Cureton has spent fifteen years as a food systems change catalyst, since 2019 leading the commercialization, adoption, and scaling efforts for new perennial and winter annual crops under development by the University of Minnesota Forever Green Initiative. In 2023, he became impatient with the pace of change in the face of the climate crisis and felt called to advance climate solutions more directly. He has since co-founded Overstory Ventures, a venture studio that builds impact-driven companies in regenerative agriculture, as well as its first two ventures Manto Foods, BV, and Midwest Hazelnuts, LLC. He carries with him aspirations for a better world for his son, deep gratitude for the teachings of mentors still with us and those lost, and a growing appreciation for the wisdom of natural systems.

Bill Davison has devoted his career to sustainable land management, from his early days as a land steward for The Nature Conservancy through his years of owning and operating a diverse, organic farm, to his current work in agroforestry with the Savanna Institute. He is a trusted voice in the agricultural community and has a talent for discussing sustainable farming techniques even among the most conventional farmers. Bill also writes *Easy by Nature*, a popular Substack newsletter about birds, gardening, and rewilding your yard and spirit.

Lee DeHaan has been leading efforts to develop perennial grains at The Land Institute since 2001. He has worked with a number of perennial grain candidates, including perennial wheat, perennial rye, and perennial chickpea, but since 2010 he has focused on the domestication of intermediate wheatgrass, to be sold as the perennial grain Kernza. DeHaan has MS and PhD degrees from the University of Minnesota, where his studies included agroecology, agronomy, and plant breeding.

Beth Dooley is a James Beard Award–winning food writer, cookbook author, and journalist and an Endowed Chair at the Minnesota Institute for Sustainable Agriculture. Her work appears in *The New York Times*, *Star Tribune*, and Minnesota Public Radio. As a writer, mom, and chef, she shares information and recipes for using unfamiliar regeneratively grown ingredients in home kitchens.

Elsie M. DuBray (Oóhenuŋpa Lakxóta/Mandan/Hidatsa) is an enrolled member of the Cheyenne River Sioux Tribe and hails from a Buffalo ranch on her reservation in South Dakota. She holds an MS in community health and prevention research and a BS in human biology with honors in comparative studies in race and ethnicity, both from Stanford University. Elsie's work focuses on Buffalo restoration, emphasizing a deep commitment to the healing potential of a restored relationship to the Buffalo for Lakxóta people and other Buffalo nations. She challenges mainstream definitions of "food sovereignty," offering a more holistic and nuanced perspective rooted in intergenerational knowledge and her family's lived experiences. Elsie continues her Buffalo work through hands-on demonstrations at her family's ranch, public speaking, and an ongoing Tribal Food Systems fellowship project.

Dr. Wendy Makoons Geniusz is a Bear Clan woman of Cree descent and a citizen of the Red River Métis Nation. One of her grandmothers was from Opaskwayak, a Cree reserve in Manitoba, and one of her grandmothers was from Poland. Geniusz was raised in Milwaukee, Wisconsin, with Ojibwe language and culture, and she has taught Ojibwe language for over two decades. Geniusz is the author of *Our Knowledge Is Not Primitive: Decolonizing Botanical Anishinaabe Teachings* and the editor of *Plants Have So Much to Give Us, All We Have to Do Is Ask: Anishinaabe Botanical Teachings*, which was written by her late mother, Mary Siisip Geniusz.

Mariah Gladstone is a Blackfeet and Cherokee woman living on her Blackfeet homeland in northwestern Montana, where she runs her business called Indigikitchen, dedicated to re-Indigenizing our food systems. She holds a BS in environmental engineering from Columbia University and an MPS in environmental science from SUNY College of Environmental Science & Forestry.

Eliza Greenman is a horticulturist and agroforestry practitioner, celebrated for her innovative work with fruit trees and sustainable agriculture. With a background in forestry and a deep passion for ecological practices, Greenman has dedicated her career to promoting the cultivation of heritage fruit varieties and advancing organic farming techniques. Her hands-on approach and extensive knowledge have made her a sought-after speaker and consultant, and her projects have helped to restore biodiversity and enhance food security in diverse communities. Greenman's commitment to preserving traditional fruit-growing methods while integrating modern sustainability principles has garnered widespread acclaim.

Rosemary W. Griffin is a native Atlantan. She spent many years working in corporate America before becoming a realtor. She is the volunteer garden manager of the Urban Food Forest at Browns Mill Community Garden.

Reginaldo Haslett-Marroquin began working on economic development projects with Indigenous Guatemalan communities in 1988. He served as a consultant for the United Nations Development Program's Bureau for Latin America and an advisor to the World Council of Indigenous Peoples and was a founding member of the Fair Trade Federation in 1994. Since migrating to the United States in 1992, Regi has led the creation and launch of multiple enterprises. He is the founder of Peace Coffee, co-founder and CEO of Tree-Range® Farms, founder and director of strategic planning and innovation at the Regenerative Agriculture Alliance, and author of *In the Shadow of Green Man*. A native Guatemalan, Regi received his agronomy degree from the Escuela Nacional Central de Agricultura, studied at the Universidad de San Carlos in Guatemala, and graduated from Augsburg University in Minnesota. He currently lives in Northfield, Minnesota, with his wife, Amy, and they own and operate Salvatierra Farms, a 63-acre farm where he continues to do to R&D on regenerative poultry production systems.

Fred Iutzi is director of agroforestry innovation at the Savanna Institute and former president of The Land Institute. He lives in Bloomington, Illinois, with his wife, Melissa Calvillo, and children, Phillip and Maria Iutzi. Back in Hancock County, his brother and sister-in-law, Russell and Jane Iutzi, watch over the Lambert farm.

Wendy Johnson owns and operates Jóia Food & Fiber Farm, located in northeast Iowa, a former corn and soybean farm turned perennial farm that integrates sheep, cattle, perennial forages, Kernza grain, native hardwood trees, fruit and nut trees, shrubs, poultry, and pigs to produce food and fiber for people. Wendy and her husband sell their goods locally and regionally and partner with other perennial farmers to help advance broader-scale adoption of local food purchasing. They also work for their family farm, a conventional corn and soybean farm. Wendy is federal agriculture policy co-lead with Climate Land Leaders and is active on several boards and committees helping further perennialize the Upper Midwest.

Megan Kaminski is a poet and professor of creative writing and environmental studies at the University of Kansas. She is the author of three books of poetry, most recently *Gentlewomen*, and two artist's books, *Prairie Divination* and *Quietly Between*. Her place-based sound, poetry, and art installations have appeared at museums, public gardens, and libraries across the country, and her poetry and essays regularly appear in literary magazines and journals.

Keefe Keeley is executive director of the Savanna Institute, a nonprofit organization advancing perennial and diverse agriculture through agroforestry. His writing has appeared in *The New Farmer's Almanac*, *Sustainability*, the *Kickapoo Free Press*, *Agri-Pulse*, *The International Journal of Illich Studies*, and *The Driftless Reader*, which he co-edited. He lived on farms in Scotland, Zambia, New Zealand, India, and Japan during a Watson fellowship year and holds a BA from Swarthmore College as well as an MS and PhD from the University of Wisconsin.

Kelly Kindscher is an ethnobotanist and prairie ecologist at the University of Kansas, where he conducts research at the Kansas Biological Survey and Center for Ecological Research and teaches ethnobotany and other classes in the Environmental Studies Program. He is the author of several books and is a passionate conservationist. He delights in overseeing three gardens, including his own, where he grows hopniss and many other edible foods and flowers.

Piyush Labhsetwar works as a land stewardship manager at Grow Food Northampton, a 121-acre community farm in Northampton, Massachusetts. He lives a mile away from this community farm with his partner. Previously, he worked as a research associate for the perennial wheat project at The Land Institute from 2017 to 2021. Piyush grew up in Central India before moving to the United States for graduate school in biophysics before he transitioned into agriculture.

Rosalyn LaPier is an award-winning Indigenous writer, environmental historian, and ethnobotanist. They work within Indigenous communities to revitalize traditional ecological knowledge and to strengthen public policy for Indigenous languages. They are the author of *Invisible Reality: Storytellers, Storytakers and the Supernatural World of the Blackfeet.* Rosalyn is an enrolled member of the Blackfeet Tribe of Montana and Métis.

Emmy Lingscheit is a visual artist working in printmaking, installation, comics, and zines. Across recent bodies of work, she explores our entanglements with the nonhuman world: interdependencies between beings and across time, from the microscopic up to the scale of the global economy. She exhibits widely and is currently an associate professor at the University of Illinois, Urbana–Champaign.

Gary Paul Nabhan is a desert agroecologist, ethnobotanist, and literary naturalist who has published over thirty-five books and 150 scientific journal articles on the interactions between plants, microbes, people, and other animals. A Lebanese American and Ecumenical Franciscan

Brother, he is currently spearheading the Sacred Plants Biocultural Recovery Initiative along the US–Mexico borderlands and around sacred springs in the Middle East, where his family is from. He has been honored by the Takreem Foundation for affirming Arab contributions to food and agriculture in the world at large and by Slow Food, Chefs Collaborative, EcoFarm, the Vavilov Institute, and Society for Conservation Biology. He grows over fifty species of agaves, two dozen perennial spice crops, and 100 varieties of fruit crops in his torture orchard near the US–Mexico border.

Dr. Pheonah Nabukalu is a plant breeder specializing in the development of diploid perennial sorghum germplasm. She conducts multienvironmental trials in Tifton, Georgia, Salina, Kansas, and Uganda, with the goal of advancing food security and sustainability through innovative breeding techniques.

Jesse Nathan's debut collection of poems, *Eggtooth*, won the 2024 New Writers Award and the Housatonic Poetry Prize and was a finalist for the Nossrat Yassini and Golden Poppy awards in poetry. His poems appear in the *Paris Review*, the *New York Review of Books*, and the *Best American Poetry*. Nathan, a lecturer in English at UC Berkeley, was born in northern California and grew up on a wheat farm in rural south-central Kansas, among his mother's Mennonite people.

Laura Paine lives, works, and farms on 82 acres of grassland in southern Wisconsin. She has bachelor's and master's degrees in plant science and more than thirty years of experience doing research, education, and market development work on grazing systems in the Upper Midwest. The current focus of her work is training the next generation of farmers and ag professionals in managed grazing.

Leah Penniman is founding co–executive director of farm operations at Soul Fire Farm in Grafton, New York, an Afro-Indigenous farm that works toward food and land justice. Her books, *Farming While Black* and *Black Earth Wisdom*, are love songs for the land and her people. More at www.soulfirefarm.org.

Ivette Perfecto is the Bunyan Bryant Collegiate Professor of Environmental Justice in the School for Environment and Sustainability at the University of Michigan. She studies the relationships between biodiversity and agriculture in an increasingly globalized economy. Her ongoing work examines the benefits of insects in rural and urban agriculture, the ecology of coffee agroforestry systems, and, more broadly, the links between small-scale diverse farming systems, biodiversity, and food sovereignty. Perfecto is an elected member of the National Academy of Science and the American Academy of Arts and Sciences and has been a member of Science for the People and the New World Agriculture and Ecology Group for more than thirty years.

Valentín Picasso is from Montevideo, Uruguay, where he studied agricultural engineering at the Universidad de la Republica. He earned a PhD in sustainable agriculture from Iowa State University. He researches forages, livestock grazing systems, agroecology, resilience to climate change, and perennial grain polyculture systems. He is a professor of cropping systems at the Swedish University of Agricultural Sciences in Uppsala, Sweden.

Dr. Muhammet Şakiroğlu is the director of natural science research at The Land Institute. Born in Eastern Türkiye, he earned his BS in biology at Harran University and was awarded the National Education Scholarship of Turkey for graduate work in the United States. He holds an MS from Iowa State University and a PhD from the University of Georgia, both in plant breeding and genetics, and he held previous faculty positions at Kafkas University and Adana Alparslan Türkeş Science and Technology University.

Dr. Lisa Schulte Moore is a distinguished professor at Iowa State University and co-owner of her family farm near Strum, Wisconsin. Schulte Moore works in large transdisciplinary teams to foster continuous living cover on agricultural landscapes, such as through prairie strips. Her research supports the development of new markets to meet societal goals for sustainable food, energy, and materials, rural prosperity, healthy soil, clean water, abundant wildlife, and wonder.

Jesse Smith is the director of land stewardship at White Buffalo Land Trust, a nonprofit dedicated to advancing regenerative agriculture. Raised on California's Central Coast, he focuses on fostering diverse partnerships for multistakeholder projects. Jesse is particularly interested in multistrata agroforestry, hydrological restoration, and specialized agroforestry crops such as agave, elderberry, and mushroom farming. He actively works to restore ecological balance in food, fiber, and medicine production, emphasizing soil health, water cycles, and biodiversity. In his role, Jesse manages the Center for Regenerative Agriculture at Jalama Canyon Ranch, a 1,000-acre site that integrates farming, ranching, and conservation.

Paige Stanley is a rangeland soil biogeochemist working to understand how grazing management influences ecosystem and soil processes that govern soil carbon sequestration and stabilization. Drawing on her interdisciplinary training—with degrees in biology and economics (BS, Georgia College), animal science (MS, Michigan State University), environmental science (PhD, UC Berkeley), and a postdoc in soil and crop science (Colorado State University)—she approaches her research through a combined lens of soil biogeochemistry, rangeland ecology, rancher sociology, animal science, and plant ecophysiology. Although understanding soil carbon outcomes from grazing management is the main pillar of her work, she also engages in synergistic projects to optimize soil carbon measurement methods, assess drivers and barriers to adaptive grazing adoption, and inform policy for more resilient rangelands under climate change.

Omar Tesdell, PhD, is associate professor in the Department of Geography at Birzeit University in Palestine. His research works to make more climate-adapted, resilient, and just agricultural landscapes. He holds a PhD in geography and sustainable agriculture from the University of Minnesota. He also directs Ardeea, an independent research group in Palestine.

Laura van der Pol is a *Homo sapiens* mom dwelling on a former prairie in North America with her wife, child, and canine companion. She works as the lead soil ecologist at The Land Institute, where she studies interactions between perennial grain crops and soils. She struggles with her deep reliance on distant, tropical trees to function and is coming to terms with living impossibly far from mountains, forests, and oceans. She finds joy in movement, a thoughtfully organized spreadsheet, and heartfelt hugs.

David Van Tassel fell in love with green creatures at an early age and is happy looking at or trying to cultivate almost any of them, from pond algae and moss to houseplants, bamboo, and crops. By training he is a botanist and cell and molecular biologist with a PhD from the University of California, Davis. He had the good fortune to learn plant breeding from Stan Cox, Lee DeHaan, and others at The Land

Institute. As perhaps the first person to try growing *Silphium integri-folium* at a horticultural scale, he is grateful for the rare opportunity to participate in domestication from scratch, made possible by Land Institute founders, colleagues, and supporters.

Gwen Nell Westerman has lived her entire life on the tallgrass prairie from Oklahoma and Kansas to southern Minnesota. Her poetry and visual art are inspired by the open landscape she loves and by the stories she heard from her grandparents and extended relatives, usually around a kitchen table. She is especially grateful for the traditional stories and plant knowledge of her Dakota people that taught her the prairie is our relative.

Index